PORTALS TO A NEW REALITY

PORTALS TO A NEW REALITY

Five Experiments to Unlock the Future of Physics

VLATKO VEDRAL

ALLEN LANE
an imprint of
PENGUIN BOOKS

ALLEN LANE

UK | USA | Canada | Ireland | Australia
India | New Zealand | South Africa

Allen Lane is part of the Penguin Random House group of companies
whose addresses can be found at global.penguinrandomhouse.com.

Penguin Random House UK
One Embassy Gardens, 8 Viaduct Gardens, London SW11 7BW

penguin.co.uk

Penguin
Random House
UK

First published in the United States of America by Basic Books 2025
First published in Great Britain by Allen Lane 2025
001

Printed and bound in Great Britain by Clays Ltd, Elcograf S.p.A.

The authorized representative in the EEA is Penguin Random House Ireland,
Morrison Chambers, 32 Nassau Street, Dublin D02 YH68

A CIP catalogue record for this book is available from the British Library

ISBN: 978-0-241-80305-9

Penguin Random House is committed to a sustainable future
for our business, our readers and our planet. This book is made from
Forest Stewardship Council® certified paper.

MIX
Paper | Supporting
responsible forestry
FSC
www.fsc.org   FSC® C018179

Dedicated to Chiara and Anna

CONTENTS

Chapter 1

INTRODUCTION

The View from Here

For more than a hundred years, physics has been treading along the paths set by two theories—quantum mechanics and general relativity—and, let's face it, it's getting pretty boring: laboratory-scale experiments keep confirming the theories we know and love. Discrepancies that might appear at the cosmological level—if there are any—are extremely difficult to probe experimentally. Unlike physicists in 1900, who were confronted with major inconsistencies between theory and experimental evidence, physicists today seem stuck. In his famous Friday Evening Discourse at the Royal Institution in 1900, Lord Kelvin talked about "two clouds" on the physics horizon, a pair of notorious problems that the best physics of the era could not resolve.* The first cloud dissipated with the discovery of special relativity (which was followed by general relativity a decade later), while the second one owes its disappearance to quantum mechanics.

You might ask, where are the clouds on the physics horizon now? General relativity and quantum physics have withstood the test of time and have shown no experimental deviations for over a century. But their successes are, I believe, coming to an end. Although there are not yet experimental discrepancies, that's not to say there are no issues. Seen from my perspective—that of a quantum information theorist—there is much to learn about some current problems in our fundamental theories of physics. It turns out that new tools from information theory may be the key to discovering physics beyond quantum physics and general relativity.

Information theory has been married fruitfully to quantum physics since the advent of quantum information theory in the 1970s. The most interesting aspect of that theoretical revolution, which led to the wave of quantum technologies we are witnessing now, is that it stemmed from an attempt to understand the fundamental aspects of quantum physics and of computer science. For my purposes here, you can think of quantum information theory as a set of concepts and methods that (1) allowed us to solve issues at the foundations of quantum theory itself (such as the measurement problem, which I will discuss) and (2) gave us a filtering lens that helped us extract and illuminate quantum theory's fundamental principles without all the unnecessary details of particular physical systems that embody this or that quantum effect.

---

* The two clouds were the relative motion of the ether with respect to massive objects and the Maxwell-Boltzmann theorem on the equipartition of energy.

Quantum information theory opens up radically new avenues for experiments that we haven't capitalized on yet. Here, I will harness all of the tools and methods we've learned about over the last couple of decades to argue that the next revolutionary step might be closer than we think.

As we will explore in the following pages, there are a handful of key hypothetical experiments that will provide us with important clues about what comes next. These experiments scrutinize elements of our theories that are particularly problematic. These elements are not just small clouds on the horizon. Rather, they lead us to the right portals beyond which lies the future of physics. Although at present each experiment is still at the level of an idea or a thought experiment, nothing is in the way of actually realizing them other than a huge inertia that makes us less likely to attempt ambitious new experiments. We simply need to take the paths that lead to these portals, and so far those paths have been little trodden.

UNNECESSARY STUMBLING BLOCKS

A major thread throughout this book is that when it comes to quantum physics, many physicists are caught on stumbling blocks that are, quite frankly, entirely avoidable. These stumbling blocks simply amount to refusing to accept quantum physics as a universal theory.

We'll unpack what exactly this means throughout this book, but here's a first example: the prevailing dogma in quantum physics, the so-called Copenhagen interpretation, contains a great number of misconceptions about what the world is made of and how to understand its most fundamental processes. First of all, it emphasizes observers and what happens when observers observe things. This leads to mystical-sounding statements such as "observations create reality" and "reality does not exist when no one observes it." Second, in the Copenhagen view, the observers themselves are considered to obey the laws of classical mechanics (the laws of Isaac Newton that coincide with our everyday experience), rather than those of quantum mechanics. This introduces an arbitrary and ultimately false dichotomy between small systems, which are thought to fully comply with the laws of quantum physics, and macroscopic classical entities (like people), which do not. Macroscopic objects such as humans are held to merely tap into the quantum world every once in a while. Third, the Copenhagen interpretation lends itself to confusion about whether the ultimate elements of reality are waves, particles, both, or neither—a topic about which I'll have much more to say.

Such a half-classical, half-quantum view of the universe not only is logically inconsistent but also serves as an unnecessary barrier to resolving whether Albert Einstein's general relativity needs to be treated quantumly. The question of whether, let alone how, to make gravity quantum has evaded us for almost one hundred years, and it might be the biggest open problem in all of physics.

The Copenhagen interpretation has also contributed a great deal of force to the consensus in physics today that quantum physics is too "weird" to work at macroscopic scales, and that quantum effects at the macro level are not accessible by experimentation. Not only are both claims false, but applying quantum physics at *all* scales is one of the keys to future physics. This puts me at odds with many adherents of the Copenhagen interpretation, who suggest that bringing together gravity and quantum mechanics requires the world to revert back to classical physics at some large enough scale. This is problematic for all of the reasons I outlined above, and then some!

Unfortunately, most other interpretations of quantum physics are not much better in this regard. You may know that Einstein worked to discover hidden variables, and that his work and that of physicists such as Louis de Broglie and David Bohm are known as hidden-variable approaches. While they do not grant special status to observers, they all still retain a classical ontology. Then there is quantum Bayesianism, which treats quantum states like probability distributions that depend on observers. QBism, as quantum Bayesianism is affectionately known, is ultimately all about how classical observers and their interactions give rise to quantum physics. This is practically the opposite of how the world actually works.

Even the closest interpretation to reality, the many-worlds interpretation, asserts that the state of a quantum system exists across many different worlds at the same time in what we call a quantum superposition. However, this has two major problems: the superpositions-of-worlds picture still does not capture everything about a quantum state, and its sometimes state-centered view ignores the other half of quantum physics, the observables. I will explain why observables are the true underlying elements of reality. To borrow from Paul Dirac, observables are represented by q-numbers (*q* for *quantum*).

In classical physics (that is, non-quantum physics), momentum, position, and other quantities we might be interested in can be mathematically represented by so-called c-numbers (*c* for *classical*). These c-numbers have the properties that we're familiar with in the classical world: any quantity of interest has only a single value, and we can specify and measure any pair of quantities simultaneously. C-numbers obey the rules we have all been taught in school about the real numbers. They are straightforward to add, multiply, and so on. Q-numbers, on the other hand, give us the counterintuitive properties of the quantum world. Any observable—for instance, momentum—described by a q-number does *not* have a single, definite value. Instead, it typically assumes multiple values at once. Nor can we simultaneously measure to perfect accuracy any pair of observables that are described by q-numbers.

Although each interpretation throws up its own stumbling blocks, one of the key ideas to overcome them will be about measurements—how we observe the observables in the first place. As I said, many interpretations of quantum mechanics ascribe a fundamental role to observers. But in reality, quantum physics does not need observers. Drawing on quantum information, one of the crucial themes of this book is that measurements are simply quantum entanglements between two physical systems, one of which can be thought of as the system under observation and the other one as the measurement apparatus. With respect to the underlying physics, which system we label "observer" and which system we label "observed" is completely arbitrary.

One consequence of this perspective on observers and observed is that it enables us to explain why the dynamics of entangled systems is perfectly predictable and, in addition to that, why the outcomes of experiments are definitive. This ability hinges on the role of the q-numbers of the two systems, and this is where quantum information comes in. You may be familiar with Claude Shannon's classical theory of information, which describes the transmission of classical bits from a sender to a receiver through some channel. Its domain is the world of information that we are familiar with: classical bits are the binary entities that make up the world of classical computation and information, and there are no counterintuitive q-numbers to be found when describing physical systems involved in transmitting these bits from one physical system to another. Quantum theory of information, however, tells us about the meaning of q-numbers as well as the *how*, *what*, and *why* of interactions between systems that are described by q-numbers. The simplest q-numbers describe qubits, which are the quantum generalization of Shannon's classical bits. Simply put, qubits can exist simultaneously in the states 0 and 1, in any ratio whatsoever. Moreover, we will see that no matter what theory supersedes quantum mechanics and general relativity, the principles of quantum information theory must also underlie *that* theory. At present, quantum physics is underpinned by quantum information, general relativity by classical information.

We'll explore the differences between these information-theoretic principles in due time, but for now, know that these differences are the main reason quantum mechanics and general relativity are incompatible. I will explain how quantum information could be brought into general relativity and what this would imply in terms of which experimental tests to pursue and how we might need to modify general relativity. That's a big claim, but in the end it will enable us to obtain a coherent description of the universe as a whole, in which everything is made up of the same kind of entity. There will be no arbitrary distinction between the classical world and the quantum world.

With these three key ideas underlying a new view of quantum mechanics—that everything can be described by q-numbers, that the observer and the observed are interchangeable, and that the principles of quantum information theory apply at all scales—we will investigate five crucial experiments that I believe will revolutionize physics.

Taking this road that has so far been less traveled will, as the poet Robert Frost would say, make all the difference.

OTHER BLIND ALLEYS

Just as some of the most popular interpretations of quantum mechanics prevent further progress in physics by asserting that the world is actually classical at some scale, a similar issue plagues attempts at describing theories that may unify general relativity and quantum mechanics. Many of these theories implicitly reject the principles of quantum information theory and the fact that systems at *all* scales can be described in terms of q-numbers.

String theory might seem to evade this problem, as it claims that both quantum theory and general relativity can be found as special cases within some deeper mathematical formalism. Other esoteric theories attempt the same thing. But at what price has this reconciliation been achieved? Most of these mathematical approaches are too far removed from basic physics. For instance, they have no experimental consequences.

Perhaps even worse, they tend to bear far too *much* fruit! They contain not just our two basic theories but—depending on the parameters chosen—all sorts of other theories that do not reflect our universe. Once we have a theory with too many things in it, we can try to winnow them down with some anthropocentric statement, such as "The laws are the way they are because we humans are here," or perhaps we throw up our hands and take the related view that all mathematical realities are actually represented and we just happen to be in one of them. Neither of these views is testable, and they completely lack explanatory power. You might as well postulate that God made the universe just like that. Furthermore, all these very forgiving theories are the total opposite of the successful physical theories we have, which are very "tight"—a small deviation from them leads us into all sorts of contradictions.

Our problems do not just stem from misconceptions about quantum physics; there are also issues around our interpretation of general relativity. Einstein gave us a beautiful geometrical picture of gravity that we will talk about at great length. However, is relativity all about the bending of space and time? It could turn out that space and time are not even among the fundamental elements of reality. In that case, approaches to quantum gravity that insist on quantizing space and time, such as loop quantum gravity (championed by physicists such as Lee Smolin and Carlo Rovelli), may be profoundly off-target. You will read about the alternative interpretations of general relativity that are more in line with quantum physics and therefore easier to put on the same quantum information footing.

The way out of these blind alleys is illuminated by the principles of quantum information theory. These principles already indicate that quantum physics and general relativity need not necessarily be incompatible in the end, and they give us hints as to what their unification might look like.

OUR ROAD MAP TO THE PORTALS

At its heart, this book consists of five experiments treated in five chapters. Each of these core chapters will be about one experiment—the motivations behind it and how it will allow us to open a portal to the future of physics. But both before we get to those experiments and while discussing them, I will also be arguing for a novel perspective on quantum physics and the theory of gravity. So the book opens with two chapters that focus on quantum physics and general relativity and will bring readers who are new to physics up to speed. I admit that my perspective is radically different from the prevailing views that are represented in the popular press and books. The story I want to tell is actually more beautiful, engaging, and convincing than the versions we have all heard for too long.

There will be a theory that eventually supersedes quantum physics, because some experimental evidence will arise that quantum theory cannot explain, as happened with classical physics. The new theory will be more general than quantum physics and will have a similar relationship to quantum physics as quantum physics currently has to classical physics. And this new theory will decidedly *not* be a return to classical physics. The experiments in this book show where and how we might find that evidence. On our journey through those portals, we'll explore domains of reality in which testing quantum theory has never been done on a range of scales: living systems such as viruses and ourselves; artificial intelligences; cosmic systems such as gravitating masses, black holes, and even the entire universe. And once we pass through them, the last part of the book presents my own speculations about the possible nature of the new theory. In particular, I'll discuss how unifying quantum physics and gravity—that is, the micro and the macro—may help us reshape even our understanding of how inanimate and living matter are related.

THE VIEW FROM HERE

There's a famous story, probably from the Zen Buddhist tradition, that involves two junior monks in training. One day while strolling around their monastery they come upon a flag that's waving in the wind. One of them says, "Look at how beautifully the flag is waving." The other one replies, "You are wrong. The flag is not waving; the wind is blowing and the flag just follows the waving of the air molecules." Their master happens to be passing by and overhears the exchange. He says to them, "Neither of you is actually right. Neither the flag nor the wind is waving. It is your mind that's waving."

There is, of course, a grain of truth to the master's claim. Our brain, which presumably provides the hardware for our mind, is the ultimate interpreter of all the sensory stimuli that we continuously receive from the external world. But it seems to me that this might be beside the point. The fable actually illustrates how the reality of any phenomenon can always be interpreted from at least three different basic perspectives. They can be hard to discriminate between on empirical grounds, yet they do offer different pictures of reality.

Before we start looking at physics from the perspective of quantum information theory, I want to highlight one of the crucial themes of this book, reflected in the moral of the Buddhist story: interpretations matter. Each interpretation offers a unique view about which elements in our physical theories are most important and how to properly explain what is going on when we observe physical phenomena. They are also crucial when it comes to discovering new physics, since some interpretations frequently lead to dead ends, while others allow us to extend and expand our theories to cover new ground.

Newtonian physics, for example, has at least three different versions: the original one due to Newton, a subsequent one due to Joseph Lagrange, and finally one more due to William Rowan Hamilton. Empirically, they are all the same (that is, they all make the same predictions), but it is primarily Hamilton's that enabled us to come up with quantum physics. This is because it contained a crucial feature that quantum physics would prove to need, which was a mathematical way of treating particles like waves. In other words, Hamilton told a better story about how the world works, even though the predictions his version made were no different from the ones made by Newton's original formulation.

I know this may come as a surprise to you, because we are accustomed to only quantum theory having multiple interpretations. In fact, though, all phenomena we study in physics have multiple interpretations. Take the notion of time, for instance. One could say that time flows, which is why things change. According to this view, time is the cause of change. The world tomorrow will be different from the one yesterday. But another, equally respectable view is that the flow of time is nothing but the fact that things change. In other words, this interpretation says that time flows precisely because the world tomorrow will be different from the one yesterday. And, just like in the Buddhist fable, there is a third view of time: the denial of the flow of time or the change of things altogether. It is only our mind that moves its attention between events, and this gives the appearance of the flow of time and the ensuing change. Needless to say, each of these views has been advocated by some philosopher or another. Each of these views is suitable for particular contexts.

Ultimately, of course, I write this book from a physicist's perspective, not a philosopher's. The topic that concerns us here is the ultimate *scientific* view of reality. What are things really made of at the most fundamental level? Our interpretation of time matters for merging gravity with quantum theory because relativity, which no future theory of quantum gravity can ignore, is all about time.

Still, we will also ask philosophical questions in the spirit of our Buddhist monks: Is there a reality out there independently of us? Or is it just our mind that actively engages in creating our reality? Many readers will be familiar with Bishop Berkeley. He was what philosophers would call an *idealist*. He famously asked: if a tree falls in a forest and no one is around to hear it, does it make a sound? The fact that he supported the "no" answer makes him an idealist, like the master monk in the Buddhist story. The opposite position—namely, that the sound exists even if no one is there to hear it—is a philosophical position belonging to a *realist*: the world exists out there independently of us or any other being capable of perception.

This goes straight to the heart of questions pertaining to the observers and the observed that I mentioned earlier. It is often said that in quantum physics, observers create reality. A measurement in quantum physics reveals one of the many possible outcomes; prior to the measurement, there were multitudes of simultaneously existing possibilities. In that sense, it would seem that quantum physics supports Bishop Berkeley's idealism. I will argue against this. Interpreting the quantum universe in a unified and holistic manner, as I do, leads us elsewhere. Making measurements in quantum physics is like any other quantum process, nothing more. In quantum information theory, there are elements of reality that exist independently of any observer. They are also local in that there is none of what Einstein derided as "spooky action at a distance" (a concept that we will revisit)—and this is despite the phenomenon of quantum entanglement, whereby two entities can apparently be linked even when separated by enormous distances. In fact, as we will see, it is *because* of (not despite) entanglement that there is no spooky action at a distance! What's more, I'll argue that entanglement is also responsible for the existence of the classical world, because—defying all expectations—larger systems tend to entangle more with their surroundings than smaller systems do.

There are many other surprises in store, including how quantum entanglement lies behind our tests of the quantum nature of gravity, and how entanglement leads us to acknowledge parallel worlds that exist simultaneously. We will come to understand why the unobserved outcomes of unperformed measurements nevertheless carry information that can affect future measurements. Imagine that: you toss a coin and see heads, and yet somehow the outcome you didn't see, tails, can still affect things in your future. This sounds unbelievable, but that's the way things are in the multiverse of parallel quantum realities. Finally, I will try to convince you that even living systems are subject to the same laws of quantum physics.

All of this will ultimately contribute to a picture of the universe that corresponds to a view of reality exactly the opposite of Bishop Berkeley's: ultimate reality exists only when there is no one looking! This, I admit, is something of a joke—reality exists "out there" whether someone is observing it or not. But there is a kernel of truth to my joke, as observers can never capture the whole of reality during a measurement.

The fact that we can never capture all of reality when we engage in observation is at the heart of what is called the measurement problem in quantum mechanics, and I hope to convince you that there is actually no problem there at all. The whole of reality is indeed always out there, and we just navigate our way through its various branches (we will get to what this branching means). Measurements are simply your ordinary quantum mechanical interactions (which I will talk about in detail), and they have no special power to collapse things or change reality. This, too, might be news for you if you've been reading the popular press.

A scientific explanation of any phenomenon has to be mathematically consistent as well as consistent with all our physical observations. Quantum physics and the theory of general relativity certainly satisfy these criteria within their respective domains. But scientific theories are also much more than that. They have to be falsifiable, which means that there ought to exist at least one experiment that could contradict them. A scientific explanation also has to present a good story in the sense that it is logically sound and can be retold, analyzed, developed, and improved upon.

So let me tell you the story of quantum physics the way I think is the most authentic and fruitful. At the core of it will be principles of quantum information processing, and our guiding idea will be the notion that our understanding of reality cannot be based on theories that have different information-processing powers. At present, quantum physics is underpinned by quantum information, while general relativity is underpinned by classical information. This is what makes them incompatible. I will explain how quantum information could be brought into general relativity and what this would imply in terms of which experimental tests to pursue and how we might need to modify general relativity.

But before we get there, we need to start at the beginning.

Chapter 2

QUANTUM PHYSICS AND GENERAL RELATIVITY

The discovery of the quantum world was kick-started by an act of desperation at the end of the nineteenth century.

As a young student, the German physicist Max Planck had been told by his university professor that in physics, almost everything was already discovered and all that remained was to fill a few holes. This very much echoes Lord Kelvin's statement that there were only two clouds on the physics horizon; it seems to have been a consensus among the physics establishment that practically everything was understood. In his forties, Planck decided to tackle one of the niggling "minor" problems—and in the process, he inadvertently gave rise to a revolutionary new field of physics.

The problem that Planck decided to investigate was blackbody radiation. In physics, a blackbody is an object that does not reflect light. When a blackbody is heated, it gives off radiation in the form of light. This is what enables a filament in an incandescent light bulb to work. The hole to be filled, so to speak, was a strange one: classical physics predicted that a very hot blackbody should give off infinite energy at high frequencies. As this is clearly not something we observe, it was given the colorful name "ultraviolet catastrophe."

Planck devised a mathematical formula to explain why this catastrophe doesn't occur in reality, leading to the conclusion that instead of being continuous, energy comes in discrete little packages, which he called quanta. He called this an act of desperation, because he didn't know why energy should be like this. As far as he was concerned, these quanta had no obvious relevance to the real world. When Planck published this work in 1900, no one, least of all Planck, appreciated what a radical discovery it was.

Then came Albert Einstein five years later with a more dramatic proposal. From the way that radiation knocks electrons out of a metal surface, he argued that if energy could be transmitted in discrete packets, then so too could light. Rather than being a continuous wave, he proposed that light is made up of little "atoms" called photons that have definite and discrete energies corresponding to certain frequencies. Einstein would call his 1905 paper in which he proposed the concept of a photon his "only revolutionary one," and it was this work, not relativity, that won him a Nobel Prize. What made it so revolutionary was that it implied that James Clerk Maxwell's classical description of light as a continuous entity—which up until then was unchallenged—actually had to be wrong.

What Einstein did for light, his Danish contemporary Niels Bohr would do for matter. Bohr had been struggling with the fact that the laws of classical physics didn't allow atoms to exist. At the time, the atom was thought to consist of a positively charged nucleus orbited by negatively charged electrons. The major flaw with this model was that, according to classical physics, an electron whizzing around the nucleus would continue to lose energy, ultimately spiraling into the nucleus and destroying the atom. And yet matter is composed of stable atoms.

Bohr resolved this by proposing that electrons move around in discrete orbits, each of which corresponds to a definite energy, and cannot exist *between* these orbits. And when electrons jump between two orbits, they emit photons. Bohr's calculations of the frequencies of these photons agreed perfectly with the data from experiments at that time.

Then something even more surprising happened, thanks to French physicist and aristocrat Louis de Broglie. If light waves are really made up of particles, then, de Broglie argued, why not also suppose that constituents of matter (atoms, protons, electrons, and such) are also *wave-like*? He adapted Einstein's photon description of light to show that electrons, originally thought of as particles, could also behave like waves. Each electron, said de Broglie, had a wave associated with it that guided its motion.

This was a revolutionary hypothesis. Einstein loved it. The idea gave a huge momentum to new research. Soon experiments on electrons and atoms confirmed that they really are waves—and that they scatter and create interference patterns when they pass through a grating in the same way that water and light waves do.

Werner Heisenberg, one of Bohr's most brilliant students, was the first person to guess the correct rules of quantum physics. His 1925 paper was highly non-transparent to contemporaries because of the radical suggestions he made about what quantum physics entails. Heisenberg dropped the classical notion that electrons move on well-defined trajectories at well-defined velocities. He said that it was unphysical to think of electrons this way because we can never measure their trajectories. For Heisenberg, both the position and the velocity of an electron became tables of numbers (that's what he called them), and these tables contained frequencies and intensities for all possible jumps (from orbit 1 to orbit 2, from orbit 1 to orbit 3, from orbit 2 to orbit 3, and so on, including the infinitely many orbits an electron can jump between in an atom).

What Heisenberg called tables of numbers were actually mathematical entities called matrices. This has a striking consequence: in matrix multiplication, which was discovered in the nineteenth century, the product of two matrices depends on the order in which they are multiplied. In other words, in the quantum world, 3×5 is no longer necessarily equal to 5×3!

The idea that all physical matter could be described as a wave was unsettling enough, but then along came the Austrian physicist Erwin Schrödinger, who really shook things up. In 1925, after giving an enthusiastic talk about de Broglie's ideas, he was asked by one of the audience members (I am paraphrasing): "You keep talking about electrons and atoms being waves, but shouldn't they then obey a wave equation? You never mentioned a wave equation in your talk!"

Schrödinger went away skiing over the weekend and came back with a wave equation—now known as the Schrödinger equation—that electrons and atoms really do obey. It is almost the kind of equation that you'd expect water or light to obey, but with one important difference: in the Schrödinger equation, time is represented by an imaginary number (the square root of −1). The fact that time becomes imaginary with respect to space also pops up in the theory of relativity. Physicists did not know whether that was a coincidence or whether the then-new physics was trying to tell us something profound about the nature of time. Time would tell.

If this wasn't confusing enough, in 1927 Werner Heisenberg realized that quantum theory predicted a fundamental limit on how much information we could ever know about a physical system. If we knew exactly the position of a particle, we could never simultaneously measure its velocity (or momentum, which is just the mass times velocity; strictly speaking, momentum is the right quantity in general because it is also applicable to the case of particles with no mass, like photons). Classically, no such restriction existed.

Some people mistakenly think that this quantum uncertainty is simply a consequence of the practical difficulty of measuring small things like electrons. Not so. This uncertainty is a fundamental feature of the universe. Heisenberg showed that, in the quantum world, objects do not possess separate properties known as momentum and position: they carry a "mixture" of the two, one that can never be completely unraveled. There are no states of a quantum system for which the position and momentum are arbitrarily well specified. This is the crux of Heisenberg's way of doing quantum physics, and the limitation bordered on heresy to those who still believed that the natural world could be fully and entirely known.

Even today, the Heisenberg uncertainty principle, as it is now called, remains one of quantum theory's most counterintuitive predictions.

Schrödinger's way of thinking was different: we cannot describe a particle as inhabiting a fixed point in space. Instead, we can only assign a set of possibilities to all the possible positions where it could exist, and a particle settles into a specific location only once somebody takes the trouble to look at it. So long as it remains unobserved, anything consistent with quantum physics is possible. Although Schrödinger himself, and later Dirac, would show conclusively that Schrödinger's interpretation led to the same predictions as Heisenberg's, it was Heisenberg's interpretation that opened the gates to a new revolution in physics, much as the Hamiltonian interpretation of classical mechanics was more capable than the Lagrangian interpretation of leading us to quantum mechanics in the first place.

Schrödinger illustrated the apparent absurdity of the interpretation that things are quantum until observed by way of a thought experiment featuring the infamous cat that now bears his name. Schrödinger's imaginary cat is a helpless creature locked in an airtight box with a vial of poisonous gas that has a 50 percent chance of shattering. According to quantum mechanics, until the box is opened, both possible outcomes are equally probable and possible. In other words, until someone peeks inside, the cat must be both alive and dead at the same time.

The apparent paradoxical nature of this logic weighed heavily on the quantum physicists of this era, and at this point Einstein—one of the grandfathers of quantum physics—became an opponent of the field. Even those who are used to the idea feel uncomfortable about the implications of the Schrödinger equation. It means that even if we know everything about a system's initial conditions, we can't predict an outcome with certainty: a particle can be in two places at the same time, and when measured it will randomly pop up in only one of them. This is radically different from the classical world, where we can predict with certainty what will happen following a system's initial state.

Einstein's first complaint was that quantum mechanics provided no deeper description of why this happened. His second major complaint was about the physical predictions that stem from doing mathematics with the Schrödinger equation: because it is a linear differential equation, we can simply add any two of its solutions together and get a third one that is just as valid. Physically, this implies a strange phenomenon known as quantum superposition. The classic example of superposition is what happens when we send a photon at a beam splitter. Rather than sending the one photon in only one direction, the beam splitter actually sends the photon in both directions at the same time. However, if you make a measurement in one arm of the beam splitter and don't detect a photon, that means you've instantaneously created a photon in the other arm—even if the two arms were thousands of light-years apart. (A light-year is—funnily enough—not a measure of time but a measure of the distance that light travels in exactly one year.) Einstein didn't like this "spooky action at a distance" because it appeared to violate the laws of relativity, according to which nothing can travel faster than the speed of light.

The prediction—and now reality—of the beam-splitting experiment and quantum superpositions has led to huge debates about the nature of reality and what happens when

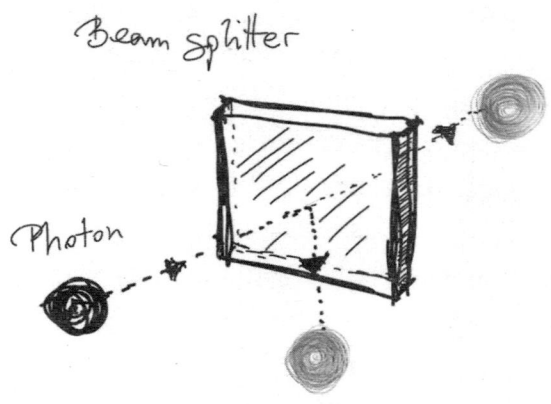

Beam splitter

Photon

A basic gadget in all experiments with photons, the beam splitter creates a superposition of the photon being in two places at the same time. The incoming photon gets reflected and transmitted by the beam splitter, which is just a piece of glass, at the same time and exists in both directions simultaneously.

we observe the universe. I will argue that these issues, too, are actually resolved, and that these resolutions tell us a great deal about what we should be focusing on as far as future experiments are concerned. In fact, I believe that progress in physics is impeded precisely because of the prevailing belief that these issues are still outstanding.

Einstein famously objected that God does not play dice, bemoaning the apparent role of randomness in the universe. I agree with Einstein, at least partially: the randomness we emphasize when we talk about quantum mechanics is there primarily because we treat the measurement process in a hybrid quantum-classical way. Namely, we think of the system we are observing as being quantum, while we think of the apparatus we use to probe the system as being classical. If, on the other hand, we think of the apparatus as being just another quantum system, randomness disappears from quantum mechanics. The quantum-classical hybrid understanding of the measurement process is what led us astray in the first place!

Isaac Newton imagined the universe to be like a giant clockwork machine running according to his immutable laws of motion: once the initial conditions are set, the universe evolves (in physics, "evolves" just means "changes over time") deterministically. This means that we can in principle predict everything that will ever happen by simply knowing only the state of the universe at the very beginning and the correct laws of motion. Admittedly, it might be difficult to obtain the knowledge of the initial state of the universe, but in principle there could be a being (imagined by French mathematician Pierre-Simon Laplace, and therefore called Laplace's demon) who could access this information. The laws of classical Newtonian physics were subject to many tests during the eighteenth and nineteenth centuries and were found to be so reliable that the Nobel Prize–winning physicist Albert Michelson famously said in 1894 that "the more important fundamental laws and facts of physical science . . . are so firmly established that the possibility of their ever being supplanted in consequence of new discoveries is exceedingly remote."

As we hinted at earlier, quantum physics has dramatically changed this picture. In quantum physics, the notion of indeterminacy enters at a fundamental level. When a quantum particle, such as a particle of light—a photon—encounters a piece of glass, such as your window, it seems to behave unpredictably. There is a chance that it will go through, but there is also a chance that it will be reflected. As far as we can tell, there is nothing in the universe that determines which alternative will happen at any given time. Surely this is randomness! But, in my opinion, the best way to think about this is

25

to acknowledge that the photon has done both; this is the true meaning of quantum superposition. The Heisenberg uncertainty principle does *not* mean that only the outcome we see is real and that the other outcome has vanished. Rather, when we measure the state of the photon, we actually become entangled with it. There is a branch of reality in which we see the photon transmitted, but there is another, equally real branch in which we (here language fails us, as this "we" is different from the first "we") see the photon reflected. And both branches exist simultaneously in a superposition. The best way to think about quantum measurements is to say that, while we cannot predict which branch of the superposition we will end up in, the superposition (now including us) remains after the measurement. In short: we join the photon in being in two states at the same time.

NEWBORN BABIES

At first glance, quantum mechanics seemed to imply the possibility of instantaneous communication over any distance, which special relativity forbids and which our experience would also suggest is impossible. The world is local and causal: local because signals between two objects must move through all the points between them, without sudden jumps, and causal because those signals move at most at the speed of light. Somehow quantum mechanics seemed to violate both with its spooky action at a distance. But this wasn't the first time that physics confronted such a bizarre implication.

In the nineteenth century, the physicist Michael Faraday used to perform a trick in his Christmas lectures at the Royal

Institution in London that seemed similarly impossible. Faraday had a large coil with a magnet at one end, and a compass at the other end, some distance away. When he put the magnet in the coil, the needle on the compass moved. To the audience, it looked like magic—like spooky action at a distance. In reality, the movement of the compass was caused by the changing magnetic field as an electric current moved through the coil, and the propagation of that field ultimately gave a completely local and causal account of Faraday's induction.

In quantum mechanics, on the other hand, there is no such explanation if we insist that the propagating entities behave classically. In the standard view of quantum mechanics, a measurement in one place (the equivalent of moving the magnet in Faraday's case) collapses something in another location, randomly and without any reason (like the needle assuming a definite direction at a distance). There's seemingly nothing happening in between the two places, no equivalent of electrons moving down Faraday's coil or electromagnetic waves propagating around it.

But this is not what I think quantum physics is telling us. When Heisenberg first began working with q-numbers, he decided to keep the classical laws of motion intact. That was because quantum physics must still recover classical physics when we encounter macroscopic objects. One consequence of the q-number formalism was the famous uncertainty principle, as we've seen, but there was more: we can apply Heisenberg's logic to everything, not just atomic physics. You want to quantize the electromagnetic field? No problem. Keep Maxwell's equations that describe classical electromagnetic phenomena, but upgrade the electric and the magnetic fields from c-numbers to q-numbers. It then automatically follows that certain pairs of components of the electric and the magnetic fields cannot be simultaneously measured.

The ingenuity of Heisenberg's approach lies in the fact that by quantizing this way, you get to keep all the good stuff from classical physics. For instance, by keeping Maxwell's equations, you retain all the desired properties of classical fields such as locality, causality, and the general compliance with (special) relativity. It is just that the things subject to these constraints are now q-numbers. Heisenberg's idea was revolutionary, but the approach in physics has actually always been the same: keep the good stuff, and modify, discard, or upgrade the bad stuff. This is not to say that there are no problems with Heisenberg's approach. For instance, there is still the question of which classical numbers we promote to q-numbers and how we do that. The short answer is: Heisenberg's approach is a work in progress.

Concluding the story about Faraday, there is an amusing anecdote related to how his lecture at the Royal Institution ended. At the end of the lecture, someone from the audience asked: "Mr. Faraday, the behavior of the magnet and the coil of wire was interesting, but of what possible use can it be?" Faraday answered politely, "Sir, of what use is a newborn baby?"

Now, the modern world of electricity owes everything to applications of Faraday's baby. But—and this is not much appreciated outside of the physics community—the modern view of gravity also owes it all to Faraday. In order to understand why, we now turn to general relativity, possibly the biggest single-handed achievement in the whole of science.

General relativity ended up being a theory of gravity, but one could almost say that this was an accident. General relativity, as the name suggests, came out of Einstein's vision that special relativity—which he discovered in 1905, the same year he published his revolutionary paper on the photon—needed to be generalized.

Special relativity only captures laws of physics for observers that travel at a constant speed. Such observers are called *inertial* observers. For example, when you are on a train that travels at a constant speed (ignoring any rail bumpiness, the wind, and so on), you are an inertial observer. We all know that unless you look out the window, you won't even notice you're moving. And when you are sitting in a stopped train at the station and the train next to yours starts moving, it is easy to get tricked into thinking that your train is moving. In other words, motion is relative—each train is indeed moving with respect to the other one, but only one of the two is moving relative to the platform.

Special relativity has nothing to do with gravity per se. It stipulates that for all inertial observers, the laws of physics ought to be the same. So whether we are on a train moving along quickly and steadily, or on a plane going ten times that speed, no physical experiment whatsoever can detect either uniform motion. We get up from our seat with no more effort than if we are not moving; a cup of coffee on a tray remains still, just as when we sit in our local Starbucks.

However, none of this explains the exact consequences of the word *relativity*. In physics, this word indicates that the measurements of spatial distances and times are different for observers moving at different speeds. Space and time are relative to the observer, and they appear different to another observer moving at a different speed. Still, the laws of physics are the same for both observers.

Now, Einstein was dissatisfied with the fact that this principle of relativity only applied to uniform motion at constant speed. Shouldn't the laws of physics be the same for all observers, even the ones that are accelerating as they jump up and down on a trampoline? In other words, Einstein thought that Nature would be illogical to make uniformly moving observers special. Either everything is relative for all possible observers doing all sorts of things, or nothing is relative at all. That's the kind of guy Einstein was—he was no in-betweener. So Einstein came up with the principle of general covariance, which holds that the laws of physics should be the same for *all* observers.

You can see that Einstein loved to think in a principled way, and his general relativity is a shining example of how far principles can take you. However, it is worth emphasizing that many of his guiding ideas, intuitions, and principles ultimately proved incoherent, incomplete, or even wrong. Amazingly, that didn't matter, because once he reached the final formulation of his theory, that was all that counted as far as physics is concerned. As the philosopher Ludwig Wittgenstein famously said, "One must, so to speak, throw away the ladder after one has climbed up it." Einstein, like many other great scientists, was a "sleepwalker." This term was coined by Arthur Koestler to stress that scientists are neither fully aware of what guides their research (frequently philosophical or psychological prejudices) nor fully aware of the implications of what they discover. The Russian physicist Vladimir Fock had this to say about Einstein the sleepwalker: "However paradoxical this may seem, Einstein, himself the author of the theory, showed such a lack of understanding when he named his theory and when in his discussions he stressed the word 'general relativity,' not seeing that in the new theory he had created, the notion of relativity was not among the concepts subjected to generalization."

This might sound like a harsh criticism; however, it is exactly what made Einstein a great physicist. Needless to say, much like with quantum physics, general relativity is subject to many interpretations, and researchers still argue vigorously about what it tells us about the universe we live in. This includes many of Einstein's principles, on which the jury is still out. Nevertheless, the equations of general relativity have so far been experimentally confirmed. They are as rock solid in their own macro domain as quantum physics is in its own micro domain.

The story of how Einstein managed such an achievement begins not with special relativity but rather with the problems that Einstein was trying to solve. In the nineteenth century, physicists had been trying to understand how light propagated. All waves, the logic went, propagate through some medium: a sound wave propagates through air or water or solids, because each of these media has atoms, and it is the vibrations of atoms that make up the sound waves. Similarly, water waves need water. So it's natural that everyone thought that light must be the same. The supposed medium in which it propagated was called the ether. However, experiments in the late nineteenth and the early twentieth century could not detect any such thing.*

---

* The speed of light would depend on the direction of propagation in the ether, but the famous experiments of Michelson and Morley never found any such variation. Much as driving into the wind slows a car down, physicists expected that the friction against ether would slow light down. But it didn't.

To make things worse, there was an apparent discordance between Newton's laws describing particles and Maxwell's equations describing electromagnetic waves. Newton's laws stayed the same when you changed position and time in a specific way, provided you applied so-called Galilean transformations between the two sets of positions and times. However, these same transformations would alter Maxwell's equations! This is highly problematic, because particles that obey Newton's laws can (and do) produce waves that obey Maxwell's equations. One way to think about such transformations of space and time is as different observers' accounts of the same experiment. Surely we expect the same experiment to lead to the same outcomes no matter how it is described. You may say that the TV is five meters away from you, while for me it's three meters away. However, this should not affect our assessment of whether the TV is on—both of us should agree, regardless of the different spatial coordinates we assign to it. It would therefore be very weird if, to each of us, Newton's laws and Maxwell's equations looked as they ought to look, while for another person—moving with respect to us—it seemed that either Newton's laws or Maxwell's equations had changed.

Einstein and others were seeking consistency. Physicists simply don't like halfway measures; either everything transforms like Newton's laws or everything transforms like Maxwell's equations. Einstein decided to side against Newton and with Maxwell.* The key to this choice was the speed of

---

*   You may well wonder why the transformations of Maxwell's equations turn out to be correct. Well, they don't in general. General transformation would change Maxwell's equations. They stay the same only for observers traveling at constant velocities. That's precisely why Einstein wanted to generalize special relativity, in order to include all possible observers.

light, which for Maxwell was the same for all observers, while for Newton this was not the case. And this led Einstein to an astonishing realization: the concept of the simultaneity of events is relative! This is the main reason Einstein decided to call it the theory of relativity: different observers do not agree on which things happen at the same time.

One might conclude, then, that observers are fundamental to the laws of special relativity. But this is not true at all; describing the experiences of human observers is just a colorful way of dramatizing what happens to the laws of physics when particles are in motion with respect to each other. That observers are superfluous is good news, since I will also argue that quantum physics does not need observers, either. As I have said, the division between the observed and the observers is arbitrary and artificial. Either one, the observer or the observed, could be considered the other.

Among the signature effects of special relativity's description of simultaneity is the famous twin paradox. Imagine one twin staying on Earth for ten years, while another twin takes a journey to a nearby star that is four light-years away. This twin travels at four-fifths of the speed of light, which means the twin on Earth waits ten years for the traveling twin to make a round trip. Remarkably, special relativity says that the traveling twin ages only six years during his trip—he is four years younger than his twin brother upon return!

Although this is called a paradox, it is not one. Moreover, we know experimentally that the effect of time slowing down for moving objects is real. Of course, we don't need human twins, conscious observers, or even living systems to test this. Any objects in relative motion to each other will do the job. We could have two atomic clocks, one stationary and one traveling away and coming back. The effect is the same. But why does a traveling clock record less time? Imagine that the two clocks emit light signals toward each other every hour according to their own times. Although I won't do the math here, the net effect of their relative movement at four-fifths the speed of light is that on the outward journey, each clock would receive a third of the signals from the other clock compared to the number of signals that it emits. However, when the traveling clock reaches its destination, it must reverse and start going back. Now, since it's going toward the stationary clock, it starts to receive three times more signals from it. So the average number of signals received by the traveling clock is now 5/3 per hour. If the stationary clock measured 10 years while waiting for the traveling clock to return, this must be equal to $3/5 \times 10$ (i.e., 6) years' worth of time for the traveling clock. Thus, less time has passed for the clock on a journey.

Sounds amazing, but let's look at this from the perspective of the stationary clock. For half of the time, it has been receiving signals from the traveling clock at the rate of one-third per hour. But when the traveling clock reverses, it is four light-years away, and the slow rate of signals continues for another four years (since the signals go at the speed of light and cannot be faster). It's only in the last year of the return that the rate jumps to three per hour. The average is therefore $((1/3 \times 9) + 3) \times 1$ year, which equals 6 years. The stationary clock also agrees that it has received six years' worth of signals from the moving clock. This had better be, since it would be weird if they didn't agree about each other's ages. That's one thing that cannot be relative! And it does not matter whether it is people, clocks, or any other physical objects experiencing it.

This irrelevance of "actual observers" is of course also true in general relativity. The word *general* means that we don't just want to describe objects moving at constant relative speeds; we want to be able to describe objects moving in any way possible with respect to other objects. It's all to do with capturing how time and space must change when one is moving in the most general possible way. And, as we've seen from Fock's disparaging statement, we should think not that special relativity is generalized but instead that gravity is added.

To understand the connection of relativity with gravity, which is as beautiful as it is surprising, we need to talk about an idea that Einstein called the happiest thought of his life (a big thing to say for a guy who had many love affairs, was married twice, and had at least three children). He imagined what a person in an elevator falling in the Earth's gravitational field would experience in his vicinity. Or, crucially, what he wouldn't experience: gravity itself!

That's for the simple reason that everything inside the elevator would be falling with the same acceleration—that of gravity. So if he let go of an apple, the apple would go in a straight line with respect to him in the elevator, as though no forces were present. On the other hand, relative to the vantage point of an observer outside the elevator, the apple, the elevator, and the person in the elevator would be falling down. This "happiest thought," that there is no gravity when falling "in gravity," forced Einstein to conclude that acceleration and gravity are indistinguishable. If you are in an elevator and someone is pushing it from the bottom up at an accelerating rate, you will be glued to the bottom of the elevator as if you were standing on Earth's surface, held down by Earth's gravity.*

The full power of this idea becomes clear when we think about something that we all find counterintuitive: that all objects, no matter how massive, fall at the same rate under gravity. Every kid makes this mistake. It seems obvious that a hammer dropped from the same height as a feather ought to reach the ground faster since it is heavier than the feather. But if you remove the effects of air—as astronauts once did on the Moon—you see that they fall at exactly the same rate. Einstein's insight explains this because an object falling toward the ground in gravity is actually the same as an object at rest with the ground accelerating toward it. If the whole ground is accelerating, then it doesn't matter what the mass of any stationary object is; the ground will reach the hammer and the feather at the same time. That's it. So, to summarize, gravity affects everything the same way, and the acceleration due to gravity is indistinguishable from any other acceleration due to any other cause (physicists call this the equivalence principle).

There is a stunningly simple experiment you can make to convince yourself that falling objects feel no gravity (and I don't mean jumping out of the window). All you need is a paper cup,

---

* There is an important subtlety here. The argument about a person falling in the elevator works fully only if the elevator is infinitely small—a point, really. This is because if the elevator is falling in a gravitational field, objects in the elevator would be falling toward the center of this gravity. So if your elevator has a spatial extent and you put two apples on opposite sides of the elevator, they would actually start accelerating toward each other. This is because each would be attracted toward the center of the earth. In that sense, if things are not all at the same point, gravity and acceleration will have different effects on distant objects. When you introduce miniature falling elevators at every point in space, what remains between these elevators, what discriminates their behavior from acceleration, is Einstein's conception of gravity.

a rubber band, a rubber ball, and some glue. Glue the ball to the band and glue the other end of the band to the bottom of the cup. If the band is long enough, the ball will be hanging out of the cup, suspended due to the balance between gravity that pulls the ball down and the spring tension that pulls the ball in the opposite direction. Now let this "toy" drop a few feet before you catch it. As soon as you drop it, the ball will end up being pulled into the cup by the tension in the band! This is because falling down actually "eliminates" the force of gravity, which means there is no force to balance the tension in the band, which therefore simply pulls on the ball and brings it into the cup.

Funny stuff this is, but the message is that in order to make gravity disappear, we should actually move without resisting it. Going with the gravitational flow makes gravity go away. Of course, we just have to make sure that there is nothing else in

A simple way to test the fact that freely falling objects feel no gravity. When the cup is dropped, the ball is pulled inside by the tension in the rubber band.

the way, like the ground (in which case we get killed by the electromagnetic force, not the gravitational force).

The upshot of all this was that Einstein solved an uncomfortable issue that troubled Newton's version of gravity: how did one planet attract another one across the vastness of empty space? Newton didn't have an answer, as his version of gravity was described by another spooky-action theory. But Einstein simply said that the first planet causes ripples in space and time in its vicinity. These initial ripples then propagate through spacetime to other locations and affect faraway objects in this way. The ripples in spacetime are called gravitational waves and were detected in 2015, leading to three Nobel Prizes in 2017. So we know that gravity propagates in much the same way as electromagnetic interactions do. The difference is that in the latter case, it is the electric and magnetic fields that wave and cause propagating ripples, while with gravity it is spacetime itself that waves. That is amazing. It's as if when you wind your clock back the resulting kink in time can propagate to other clocks, which themselves then start to feel the lag.

But that's not the only way to interpret gravity. You might hear it sometimes said that gravity causes clocks to run slow. As popularized in movies like *Interstellar*, spending a short amount of time near a massive object like a neutron star could amount to a lifetime here on Earth. However, it is much more accurate and closer to the spirit of Einstein's original idea to say not that gravity is the cause of this difference in time but instead that gravity is the difference itself. One, after all, ought to follow William of Occam, the famous medieval philosopher, and not multiply causes beyond necessity. Ultimately, one could say that gravity is the variation in how time flows from one spatial point to another.

WHY THINGS FALL,
ACCORDING TO EINSTEIN

Here is an appropriate puzzle, which is really more like a joke. A guy wakes up in the middle of nowhere in a tent. He comes out of the tent only to see a bear. Fearful for his life, he starts running southward for a kilometer. Then he decides it's better to change course and goes eastward for another kilometer. Finally, he is sure another swing will do the job of escaping the bear, and he turns northward and runs for another kilometer. At the end, to his astonishment, he is back where he started and the bear is, unfortunately, still there. Question: what color is the bear?

Ha, ha, ha. You might wonder: how are the details given in the puzzle at all relevant to determining the color of the bear? But the question is only seemingly a joke. A more straightforward question is: where on Earth is the tent? Or, better still: what shape is the Earth?

The scenario outlined with the person running away from the bear is possible only because the Earth is (to a good approximation) a sphere. If you start at the North Pole, go south, then go east, then go back north, you will indeed return to the North Pole. At no other place on Earth would this work (and, yes, the bear is therefore white). Also, the question would not work if the Earth was flat, since then you would walk along the three sides of a square and would also need to walk along the last remaining side (due west) to come back to where you started. This is definitely something for the flat-earthers to ponder.

Einstein's general relativity says that gravity is the curvature of spacetime. The fact that gravity is indiscriminate, and that *anything* that has energy also gravitates, explains a multitude of phenomena that Newton never could. There is the bending of light, there is the precession of Mercury's perihelion, there is the time dilation due to gravity, there are black holes, there's the cosmological constant, and so on. It's important to emphasize that it's not just that general relativity can explain things that Newtonian gravity cannot. Any new theory of physics, even in the domain where it agrees with the old theory it has replaced, still brings along a completely different account of one and the same phenomenon. A good example is the famous falling apple that apparently led Newton to think that the Moon is pulled by the same force that originates from within the Earth. This then gave him the idea that gravity is a universal force and acts on everything. Indeed, if we asked Newton why the apple falls, he would have said that the cause is the force of gravity generated by the Earth.*

For Newton, objects stood still or traveled at a constant speed in a straight line unless they were acted on by a force. As simple as that. You take a stone and throw it as far as possible. The path of the stone will be curved: it will first climb up and then, once the highest point is reached, come down, much in the reverse manner from that of climbing up. The trajectory is a curve called a parabola in mathematics. It's definitely not a straight line, and this is for Newton a proof that a force (in this case of gravity) is acting on the stone's trajectory to give it a bend. The same is true for all other forces—they all "force" particles to deviate from a straight path, which is their natural motion in the absence of forces.

---

* To be sure, Newton was not entirely satisfied with this explanation. He did have a problem with the fact that the force of gravity apparently was not a contact force in that it acted at a distance, since the apple and Earth (or the Earth and the Moon, and so on) were separated by the space in between. In other words, he believed that there must be some mechanism that transmitted this force from the Earth to other objects. However, he didn't know what this mechanism was. As we saw, Einstein solved this apparent violation of locality.

Now, Einstein's explanation could not be more different. He says that the trajectory a thrown stone follows is still actually a straight line! How can this possibly be? It can be, says Einstein, when one includes *time* in the description. Time runs slower closer to the Earth than higher up. In Einstein's four-dimensional spacetime, a straight line is simply the "shortest" path, the one normally requiring the longest amount of time, between the two given points—say, the point where you throw the stone and the point where it finally landed. It's just that in relativity, the time is equivalent to space with a minus sign, which implies that the shortest path will involve the most time.

So, to achieve the longest time, the stone goes up as fast as possible, then slows down as it reaches the highest point, stays there for a bit, and then falls back down to come to its final point of rest. In other words, the stone "tries" (unintentionally, of course) to spend as much time in the regions (higher up) in which time ticks faster. In Einstein's spacetime, the stone thus describes a straight line in four dimensions, even though in Newton's three-dimensional space version this looks like a curve. Analogously, walking in a straight line on Earth looks to someone following your progress on a two-dimensional map like you are walking along a circle. In general relativity, going along a straight line looks like curving when projected into the three-dimensional space.

There is a very apt analogy of a lifeguard on a beach. The guard is on the beach, twenty meters away from the closest point to the seafront. Suddenly she hears a child screaming, "I am drowning—please help me!" The child is about ten meters away from the beach but diagonally away from the guard. What's the fastest way for the guard to reach the child?

The guard could run straight to the sea for twenty meters, but then she has to swim diagonally in the water—and swimming is much slower than running. This could therefore take an unnecessarily long time. A better strategy would be to run a bit longer on the beach toward the child and then swim less. Indeed, the best strategy is one such path that looks like a broken line in the two-dimensional space of the beach and the sea. However, when time is included as another dimension, the path becomes a straight line.

Einstein thought about gravity in exactly the same geometrical way. Particles are going straight, but the underlying spacetime is curved. And the reason it is curved, the reason spatial distances and times vary from space to space and from

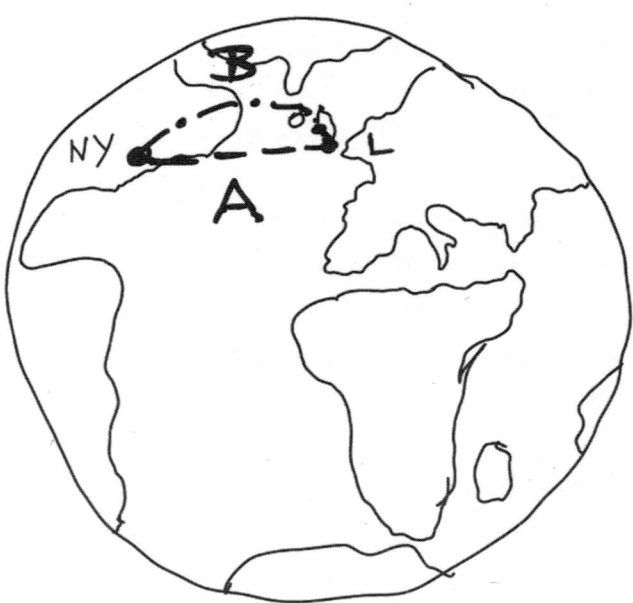

The shortest distance on a sphere such as our planet looks curved when presented on a two-dimensional map.

time to time (I know, this is complicated to imagine for me, too), is because of gravity (which is generated by anything that has energy, such as massive objects, but also even by the electromagnetic and gravitational fields).

We don't experience this curvature because it's tiny. When we are standing, the time at our waist runs 1×10^{-16} faster than at our feet. This is a difference of a ten-millionth of a billionth of a second! Our atomic clocks can and have measured this discrepancy, but it is not something that we are biologically equipped to feel (luckily, I might add, for it would be terrible to feel your head age faster than your feet).

And it's not just time that warps. Space also distorts. A cat stretched out at our feet is a ten-millionth of a billionth of a meter shorter than its twin sprawled at our waist. But this difference in length is one-tenth the width of a single neutron and is therefore clearly impossible to see.*

This warping led to a crucial objection against Einstein's work from the physicist Paul Ehrenfest. We've all been to an amusement park and have taken rides on carousels. It's frequently an exhilarating feeling to enjoy a ride in circles, not least because there is a force—the centrifugal force pushing you outward—that you need to resist in order to stay on the carousel. Special relativity says that distances contract in the direction of motion. So the circumference of the carousel wheel ought to contract from the perspective of our rider. But special relativity also says that nothing happens to the directions

---

* Speaking of which, why does the falling apple care about such small differences? Because in relativity, when time is measured in the same units as space, falling down for one second is actually worth about 300 million meters. So the curvature of time is much more important under such circumstances than the curvature of space!

perpendicular to the direction of motion. So the radius of the carousel is the same for the person riding as for the person standing by and watching the carousel.

Hang on a minute. The circumference is smaller, but the radius is the same. Any twelve-year-old, however, knows that the circumference of a circle is 2π times its radius. Einstein's special relativity is thus surely wrong, as it contradicts basic Euclidean geometry. That was Ehrenfest's objection. And special relativity is certainly wrong here, but the motion on a rotating disc involves forces, which means that it's not a uniform motion, and therefore the disc cannot be understood just with special relativity. In fact, it points to why (and how) special relativity needs to be generalized. Because spacetime is curved, it is not subject to the laws of Euclid. In non-Euclidian geometry, the circumference of a circle could be more or less than 2π times its radius!

Having learned all this, we can now understand how gravity reconciles apparently different local (and relativistic) accounts of what's happening in any particular scenario. Let's use the free fall in Earth's gravity from one height to another to illustrate the point. This is very similar to the twin paradox, except that now we need triplets. One triplet is in continuous free fall at one height from Earth's surface, another in continuous free fall at a lower height, and the third is falling from the first height to the second. Now say that the third falling triplet's speed is v_1 when passing the first triplet and v_2 when passing the second triplet. The ticking clock rates of the triplets at these two heights will differ precisely because of the different velocities, which means each of the other triplets seems to move away from the third at different speeds. And the equivalence principle says that they are all inertial observers,

44

because they are in free fall! (Remember that for Newton inertial motion is motion at constant velocity, whereas for Einstein inertial means that you experience things as though there is no gravity.) While the laws of special relativity hold locally at every height, what connects the two special relativistic frames at different heights is gravity. It is crucial that the third triplet falls under gravity—it is this that determines the two velocities that ultimately tell us about the difference between the clock ticking rates at two heights.

Of course, if you apply this to the whole of space, we get a very interesting image of what's going on. At each point, there is a freely falling clock, which—as soon as it advances a little— gets taken away and is replaced by a new freely falling clock starting from the final time recorded by the preceding clock. Luckily, just as with the fictitious observers, there are no clocks out there in space; it's just that the spacetime behaves as though there were. Needing clocks to define spacetime would run into a circular logic, since in order to construct the clocks we need to understand the dynamics in spacetime. Spacetime, or whatever else determines it, must therefore be prior to clocks.

The effects that acceleration has on time are why sometimes people say that gravity is not like other forces and that Einstein in fact eliminated gravity by replacing it with the "curvature" of spacetime. Space and time get warped together as a single entity, and this is what we call gravity. That's all well and good if we don't have any interest in quantizing gravity, but if we do, eliminating gravity doesn't seem to help the cause of making the key features of Einstein's gravity compatible with the main tenets of quantum physics. These difficulties prompt many people to claim that the two theories are irreconcilable.

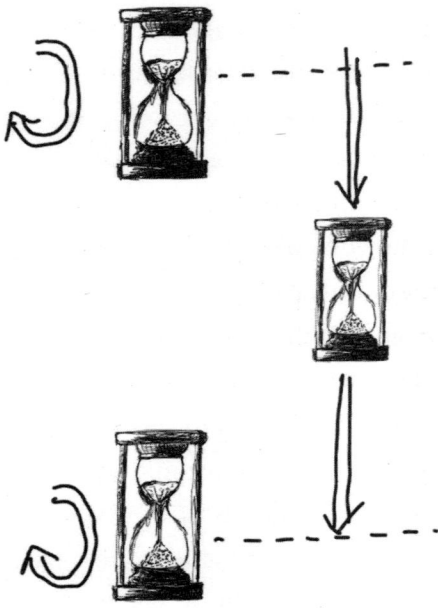

This is how general relativity accounts for free fall in gravity from one height to another. At each of two heights, there is a clock; in addition, there is a clock falling between the two heights. Einstein's universe is packed with imaginary clocks whose different ticking times account for gravity.

I want to show you how quantum information suggests to us a number of accessible experiments to probe the region where quantum physics and general relativity meet. The surprise will be that many such experiments are tabletop and can be done for a relatively small cost.

WHEN THE TWO BEST THEORIES DON'T AGREE

There have been numerous tests and confirmations of general relativity. The first three predictions proposed by Einstein were the precession of Mercury's perihelion, the bending of light rays by gravity, and the gravitational redshift. These either were already measured by the time Einstein made the prediction or were confirmed in the following years. Then came the Shapiro effect (the slowing down of radar signals near the Sun), gravitational lensing, frame dragging, the Mössbauer rotor experiment, tests of the equivalence principle, direct observation of black holes, binary pulsars, and the direct detection of gravitational waves. And the list goes on and on.

As much as we have no reason to doubt quantum physics in the micro domain, the macro domain of astronomy and cosmology complies equally well with the laws of general relativity. And yet I will here describe an experiment, accessible with current technology, to show that both cannot be right at the same time. The experiment illustrates how we are very much in the same position as Einstein was with Newton and Maxwell when he had to decide whom to side with. The only difference now is that we have to decide whether to side with Heisenberg or Einstein. And it is here that our philosophical interpretations of physics, and the principles of quantum information, will help us decide.

Briefly, if one thinks that gravity is always going to be just the classical warping of space and time, then one need not quantize gravity. One just needs to write other fields (such as the electromagnetic one) in terms of the "curved" spacetime coordinates. On the other hand, if one is convinced that gravity *is* like the electromagnetic field, then its components that look like the electric and the magnetic fields need to be made into q-numbers. Finally, if one is convinced that there is no gravity and there is just spacetime, then it is the spacetime that needs to be quantized.

One way in which philosophy matters here is that we must ask what we think of "being versus becoming." This concerns our understanding of time and uncertainty. Is reality composed of one thing after another (as Heraclitus would have it— you cannot step into the same river twice), or has everything that we think of as being in the future "already happened" in some sense and exists in a higher-dimensional, timeless reality (as Parmenides thought—change is simply an illusion)? There are all sorts of mixtures of these two extreme positions as well. Plato, for instance, thought that the things we observe are changeable, but this is only because they are poor replicas of what he thought were abstract and immutable forms that make up a kind of ultimate reality (so he tried to unify Heraclitus and Parmenides to show that you can have it both ways).

What do our two physical theories tell us about becoming and being? Quantum physics seems strongly on the side of becoming. Identically prepared experiments will yield different, unpredictable results. All the information we have about our experiments is not enough to tell us how they turn out. The experimental outcome is genuinely new, not fully conditioned on the past. Even when we have maximal information, new information always arises from observations. This is because of the Heisenberg uncertainty and seems more Heraclitean than anything Heraclitus ever said: you really can't ever step into the same quantum experiment twice.

But our other current best explanation of physical reality, general relativity, seems to support Parmenides. Hermann Weyl, one of the early aficionados of Einstein's relativity, thought that the concept of "passage of time" makes no sense as far as physics is concerned. It is simply a psychological effect. Here is what he had to say: "The objective world simply is, it does not happen. Only to the gaze of my consciousness, crawling along the lifeline of my body, does a section of this world come to life as a fleeting image in space which continuously changes in time." And Einstein himself tried to console his best friend's son and sister using relativity, saying that even though his best friend had just died, we are actually all dead with respect to some suitably chosen observer (you can judge for yourself how well this worked as a consolation, but Einstein's phrase, "the passage of time is an illusion, albeit a persistent one," is Parmenidean for sure).

This is true for any event you can think of, such as a click of a photodetector in some quantum experiment: there is an observer for whom this event lies in the past. But that means that they already know its outcome! So even if the detector clicking or not clicking lies in my future, and the outcome is completely random and therefore unknowable to me, there is some other observer in the universe for whom this event has already happened.

So suppose you are standing still looking at Jupiter (probably the brightest "star" you can see these days at night if you are geographically close to me). Suppose further that there is someone on Jupiter sending a photon through a beam splitter in their own laboratory and observing which way it comes out. Now, if the guy on Jupiter has just fired the photon, and this photon has not yet gone through the beam splitter, then which way the photon goes lies in the future for him as well as for you. The outcome is not known to either of you at that point because things are simultaneous for both of you (since you are stationary with respect to each other—I am ignoring the motion of the Earth relative to Jupiter).

However—and this is the mind-blowing bit—for a passerby whose position is the same as yours but who happens to be moving away from Jupiter, this event lies in her past! As far as she is concerned, the photon not only has been fired but also has entered and exited the beam splitter and has already been detected in one of the two detectors at the outputs.

"You've gotta be kidding me," I hear you say. "How can this be?" It can be (logically speaking; see below) and it is (factually speaking; see below). This is the main message of relativity: namely, that simultaneity is relative.

What I think of as things happening at the same time for me—I am in the bath and my phone in the living room rings—someone else might see differently. There are observers for whom the phone rings before I enter the bath, in which case

Two people walk past each other in opposite directions. Because of this, it is possible that from the perspective of one of them a distant event has already happened, while for the other person, it is yet to happen, even though both are located at the same place. This consequence of relativity can successfully be combined with quantum mechanics.

they are thinking, "Why in the world did he not answer it?" And there are observers who see it the other way round, in which case—feel free to make up your own joke at this point.

But doesn't this contradict the quantum randomness we discussed before? I mean, how can anything be random if there is always an observer who already knows the outcome? (I don't mean God; I simply mean someone moving in a certain way with respect to you.)

I'm going to pause here to say that this is why I love physics. Just look at the tightrope we have to walk to try to reconcile seemingly contradictory aspects of Nature that arise from our different theories and experiments. But reconcile we can. And it all has to do with the fact that while there is no absolute past, present, and future in relativity, there are no such things in quantum physics, either! This might come as a surprise to some of you, but it ought not to if you've understood that in quantum physics, even the unobserved outcomes of present events can affect future events. Past, in the sense of what has definitely happened, is also relative in quantum physics— relative to observers (which, as I keep emphasizing, are nothing special—they're just other physical systems).

This point is very much related to the fact that it is better to talk about quantum indeterminism or undecidability rather than randomness. Quantum mechanics isn't really random at the level of the whole universe. In fact, it is fully deterministic.

For the person on Jupiter, then, what it means to observe which detector clicks is that he becomes entangled with the state of the detectors. That entangled state has two branches. In one branch, the person on Jupiter has observed one detector click, while in the other he has observed the other detector click—and both branches exist in a quantum superposition. In each branch, the observer in that branch has seen a definitive event. But as far as another observer is concerned, both the Jupiter observer and the photon still exist in an entangled state. Just like Schrödinger's cat.

The "still" observer on Earth is simultaneous with the state of the photon before it entered a superposed state (at the output of the beam splitter), but the moving observer is simultaneous with the entangled Schrödinger's cat–like state. Therefore, a potential conflict between the randomness of becoming and the determinism of being is solved through quantum entanglement as far as quantum physics and relativity are concerned.

In relativity, locally (to an observer) things are *becoming* even though another observer can see them as already *being*! In quantum mechanics, we can have a perfectly static picture of an enormous entangled state in which all things have already happened, while in each branch they "look" to the relevant observers as becoming. All the experimental portals I have in mind that will take us to the new physics theory are underpinned by this kind of logic.

One consequence of gravity being a curvature of spacetime is that while Newton's gravity is captured by a single "component" (the gravitational acceleration, or potential, which needs to be specified at all points), in Einstein's general relativity you need ten different components.

From our perspective, the key question about the ten components of gravity is whether they ought to remain ordinary numbers, or whether—like the six components of the electromagnetic field—they ought to be upgraded to q-numbers. I think the latter will turn out to be the case, and a part of this book is about experimentally testing this idea.

The amazing fact about this is that we can and are about to test whether (some of) gravity's components need to be described by q-numbers with current technology.* The power of such a result is enormous—it would make a huge impact on physics. And it's all thanks to quantum information and entanglement.

---

* Equally amazingly, the test would show whether or not some aspects of gravity are q-numbers, even if it doesn't help us identify *which* aspects exactly are q-numbers.

Chapter 3

THE WORLD EXISTS
ONLY WHEN IT IS
NOT OBSERVED

I hear lots of complaints about there not being much progress in discovering new physics. But to me this isn't surprising, considering the vast number of practitioners of quantum physics who still believe that we have not solved the measurement problem. These scientists claim that quantum physics is still incapable of consistently describing experimental observations. Therefore, they say, it should be modified by extra mechanisms such as the "collapse of the wavefunction." As a result, much of the field is still stuck on discussing the very same issues that were raised at the Solvay conference in 1927, ignoring the progress that has since been made. In sharp contrast with these views, here I want to make the point that if we instead embrace quantum theory as a universal theory and all of its consequences, not only is the measurement problem dead and buried, but we can also clearly see a path toward discovering new physics. This will take us to our portals, and it all starts by understanding how observation is completely compatible with quantum physics.

Bishop Berkeley famously asked if things still exist when they are not perceived by observers. In his case, the observer did not need to be a human being. Berkeley was satisfied with the fact that "a tree in a quad always existed because it is observed by Yours Truly, God." His concern with existence being dependent on perception was perfectly natural. After all, we make inferences about the external world only when information from our senses is conveyed to and interpreted by our brain. So, strictly speaking, reality could all just be in our heads and may not exist outside or independently of it. This philosophical position, advocated by Berkeley himself, is known as subjective idealism.

Similar claims have often been made about quantum physics, but I'd like to now entertain the idea that quantum physics presents us with the exact opposite view of the world. I guess it would be appropriate to call it objective realism. It says that things fully exist only when they are not observed.

As we've seen, the quantum world is made up of q-numbers. These are entities that cannot be reduced to a single real number, and in this respect, quantum physics is radically different from classical physics. These q-numbers describe everything in the universe. There is a q-number, for instance, that describes the electric field at a given point in space and at a given instant of time. There is also a q-number that pertains to the magnetic field at the same point in space and at the same time. And the inherent indeterminacy of quantum physics means that when we make a measurement of the field at that point, we will be able to see only the value of the electric field or only the value of the magnetic field, not both. But this instance of the Heisenberg uncertainty principle doesn't tell us that the electric or the magnetic field quantum numbers exist unless we look at them. Rather, it tells us the opposite: that they exist independently of any observers. Nevertheless, we can see only one or the other, never both simultaneously. So the world that we see when we measure or perceive (for perception is just a kind of measurement) in our neighborhood is different from the world out there when we do not engage.

Everything in the universe is characterized by q-numbers. When different systems interact—and these systems could be something like an atom and a physicist—then the q-numbers pertaining to the atom become intertwined with the q-numbers of the physicist. This means that the atom and the physicist become quantum entangled, such that if the atom is in one of its many states—say, the excited state—then the physicist is in a corresponding state in which his brain says, "Oh, I see that the atom is now excited."

This explains why we see not q-numbers but only the individual c-numbers of everyday life. Our mental states correlate only to other *states*, not to q-numbers (which, as I've said, correspond to *observables*). Each q-number is really a collection of states, and measurements are only correlating individual states of different q-numbers. This is one of the rules of quantum mechanics. And which particular states we will correlate to is not determined, even though the overall process of interaction is deterministic. So we deterministically couple to an atom, but we cannot know in advance if we will see it in the excited state or in the not-excited state.

In fact, a better way to think about this is the other way around—namely, that the states we correlate to ultimately determine who we are. The excited atom leaves us with a memory different from the one that the not-excited atom leaves us with. It also leaves us with different emotions, since, for instance, we could have been disappointed that the atom was not excited.

When Heisenberg came up with his ingenious idea for modifying classical physics, the focus was on the atomic spectra, specifically how to properly describe the absorption and emission of light by matter. When we shine light on, say, a piece of metal, it tends to absorb some of the light, store it for a while, and then emit it back out into space. All of the aspects of this process—the absorption, the storage, and the emission—were completely mysterious and poorly understood. Heisenberg's approach was inspired by Einstein's philosophy that had led to special relativity, which was to base his theory on observable entities only. Any references to entities that cannot be directly measured ought to be purged. In atomic physics, for instance, one cannot directly measure the trajectory of the electrons inside an atom. However, one can observe the color (i.e., frequency) of the light emitted by the atoms when electrons move from an excited state to a less excited one—which is to say, when they move from one orbital to another orbital that is closer to the atomic nucleus. One can also observe the intensity of this light—that is, the number of photons emitted by atoms as their electrons transition between orbitals.

The crux is that if we want to describe the position of the electron, we need to start with two numbers: one pertaining to the orbital the electron comes from and one telling us the orbital the electron finally ends up in. Both the frequency of light emitted by the atom and its intensity ought to depend on these two numbers. That was Heisenberg's key insight.

This does not sound that momentous. In classical physics we would be talking about the position and velocity of an electron, while here we are instead talking about the frequency and intensity of light. But now comes an interesting consequence that will explain why the step Heisenberg took was truly revolutionary. The velocity of the electron is just the rate of change of its position. The velocity simply tells us how quickly in time the electron goes from one position to the next. The faster the electron goes from A to B, the higher its velocity. Since, according to Heisenberg, the electron's position depends on more than one number (namely, the two orbitals), so must its velocity. From there, Heisenberg started building tables of numbers, with rows 11, 12, 13, and so on, and columns 11, 21, 31, and so on. There was one table for the position and another table for the velocity. The numbers simply correspond to the orbitals that the electrons move between, so that "13" indicates that the electron transitions from orbital 1 to orbital 3. Mathematicians already knew about these tables; they had been invented by Hamilton (and Arthur Cayley) some fifty years before Heisenberg and were called matrices, but Heisenberg didn't know about them at the time he was discovering quantum mechanics. (They simply hadn't made it into the German undergraduate physics curriculum in 1920!) Heisenberg must have been shocked when he started to multiply them. If you multiply the position by the velocity, you won't get the same result as when you multiply the velocity by the position. In other words, the order in which matrices are multiplied matters—$M \times N$ is not the same as $N \times M$!

Classically this would make no sense at all, since the position and velocity are just single numbers: 5 meters × 4 meters per second is the same as 4 meters per second × 5 meters. However, when Heisenberg purged classical physics of the entities he thought were not observable, the most striking consequence was that the position and velocity no longer commute (to say that something does not commute is a mathematical way of telling us that the order in which we multiply them matters). This was the first glimpse of the Heisenberg uncertainty principle. Heisenberg's second great insight was that he didn't need to jettison classical physics entirely; all he had to do was change the c-numbers to q-numbers.

Let's take a very simple example to see how this works in practice. Newton says that if there are no forces acting on a particle, it moves in a straight line and at a constant velocity. This gives us the famous equation that the position of the particle is just the product of the velocity and time. If the velocity is 5 meters per second, then after 10 seconds the particle has moved 5 × 10 meters, or 50 meters. In quantum physics, particles don't go in straight lines even when there is no force acting on them. And this is simply a consequence of the fact that multiplying the position and velocity one way does not give the same result as multiplying them another way.

Here is a simple way to understand this. If multiplying things depends on the order, then dividing them must, too, since division is multiplication in reverse. Therefore, the classical equation that the distance divided by velocity is time can be run in two different ways in quantum mechanics. You could calculate 1 divided by the velocity and then multiply by distance, or you could multiply the distance by 1 and then divide by the velocity. These two ways of multiplication result in two straight lines that become farther and farther apart with time! In fact, a quantum particle can be found anywhere between these two lines, which implies that its position spreads with time.

As we said, this is a consequence of the uncertainty principle. In quantum physics, an electron starting in one position will end up in lots of different places with time, and we can only assign probabilities to it being in each of these various places. We can never tell with certainty where the particle is until we have entangled ourselves with it. A quantum particle thus behaves more like a wave—it spreads as it propagates and, like a wave, exists in many places at the same time.

This brings us back to Schrödinger. He gave us an equation—again in perfect analogy with classical mechanics—that these quantum entities obey. This is a wave equation, but what does the waving is an object that Schrödinger called the "catalogue of information" and which we now call the wavefunction. The Schrödinger equation, which is based on the classical variant of Newton's laws called the Hamilton-Jacobi equation, tells us how quantum states—which are defined by the wavefunction and which, along with observable attributes such as position and velocity, describe quantum systems—change in time. If a particle is in a superposition of being here and there, then what will this state look like in half an hour's time?

Dirac's way of doing quantum physics showed that the two approaches, Heisenberg's and Schrödinger's, are actually the same as far as their predictions are concerned. Those two approaches do, however, emphasize different aspects of quantum physics and—just as we said of different formulations of classical physics—could be more or less useful in finding the next theory of physics.

Let's now discuss a confusion that resulted immediately from the discoveries of Heisenberg and Schrödinger. It is to do with the detection of alpha particles, and in my view, it teaches us all we need to know about quantum physics.

PARTICLES AND WAVES

Newton thought that everything is made up of particles. However, various aspects of light could not, at Newton's time, be explained if light was also made up of particles. Ultimately, Maxwell—standing on the shoulders of Christiaan Huygens, Thomas Young, Augustin-Jean Fresnel, Michael Faraday, and many others—proved that light is actually an electromagnetic wave. At that point, the world (still classical) was understood to be made up of two kinds of entities, particles (constituents of matter) and waves (such as the electromagnetic field).

Then came Einstein who (seemingly, but not really) reverted back to Newton and said that light is, after all, made up of particles (photons), though he did admit that they can also behave like waves (which is how we can account for the interference of light). As we saw, de Broglie hypothesized that material quantum objects (such as electrons) can also behave like waves. This led to a unification (everything is a wave and a particle) and inspired Bohr to talk about complementarity—namely, the fact that all entities can manifest their particle-like or wave-like nature depending on how we "look at" (experiment with) them. This is at the heart of the famous wave-particle duality of quantum systems.

So, then, what is the right ontology when it comes to reality? Is the world made of particles, waves, or things that behave sometimes like particles and sometimes like waves? According to quantum physics, everything is actually made up of waves. But these are q-waves, meaning that the entities that are doing the waving are q-numbers.

If particles are waves, then the behavior of alpha particle decay in a cloud chamber seemed to present a paradox for quantum physics. An alpha particle is a helium nucleus (it's got two protons and two neutrons), and it sometimes gets ejected in the decay of a larger nucleus. A cloud chamber was a great invention to observe such particles (worthy of several Nobel Prizes), though nowadays you can make one in fifteen minutes in your own house with existing kitchen utensils (there are many YouTube videos on this if you want to try making one). The idea, as the name suggests, is to have a particle travel through a gas that can readily be ionized (i.e., a gas whose molecules have their electrons stripped away by collisions with the particle). As the alpha particle collides with the gas molecules, it ionizes them in succession. Ions are charged, and the charge attracts neighboring gas, which condenses around the ionized gas. Therefore, the traveling and colliding particle leaves a track of condensed vapor in its wake, like a jet plane flying through the sky.

So far so good, but the problem was that the tracks are always straight lines. If, as quantum physics suggests, everything is a wave, why do we get straight lines from alpha particles instead of concentric circles (like waves spreading in a pond when we throw a stone into it)?

The explanation was provided by Nevill Francis Mott in 1929. Mott said that a single particle simultaneously traverses all the tracks in all directions (as a wave would), meaning that all trajectories exist in a superposition and at the same time. It's just that when we look at it, we can see only one of these trajectories.

But we ourselves are also a collection of q-waves. Our q-waves correlate with the q-waves of the alpha particle, and these correlations between q-waves are what quantum entanglement is. And when entanglement happens, c-numbers emerge. The classical world owes its existence to quantum entanglement.*

In the case of the alpha particle, a single trajectory emerges because the q-numbers of the gas become correlated to the q-numbers of the alpha particle, and both become correlated to the q-numbers of the physicist doing the experiment. But the same would be true even if there were no humans around to observe: the collision would be described by the interaction of two q-waves (just as Mott described it). The beauty of this picture, and what I believe to be the main message of quantum physics, is that there is no division between observers and the observed. They are completely interchangeable. In terms of information, the observer has as much information about the observed as the other way around, so each could play the role of the other. Just as it could be said that we observe the trajectories of alpha particles, alpha particles could be said to be observing various different states of our perception. Everything—according to this philosophy—is captured by q-numbers. The whole universe can be seen to be a huge collection of various q-numbers.

In lectures given toward the end of his life, Schrödinger clearly spelled out the same picture that Mott argued for: that everything is a q-wave. He advocated this view not only

---

* This constitutes our most accurate description of nature, called quantum field theory. A particle in this theory is just one stable configuration of the underlying q-wave (or, in the more formal language of quantum field theory, a single excitation of the quantum field).

because it avoids the confusion arising from the dualistic wave-particle language but also because it contains no collapses of the wavefunction, no abrupt discontinuities due to measurements, and no quantum jumps (he was particularly keen to avoid quantum jumps, about which he said that if they turned out to be true, he would wish he'd become a plumber and not a physicist).

In fact, quantum physics is pretty much captured by the four nos: no collapses, no quantum jumps, no observers, no underlying c-numbers. It is interesting that Einstein complained about quantum physics precisely because he thought that the four nos were, in fact, four yeses.

How far have we tested this q-wave picture? Although no one has yet done the experiment of Schrödinger's cat, simply because bigger and more complex objects are exceedingly hard to control quantum mechanically, scientists have shown experimentally that all subatomic and atomic particles, and even some simple molecules, are capable of existing in many states at the same time. There is evidence that some more complex chemical processes are also fundamentally quantum mechanical.

And now we are starting to experiment with living systems. My colleagues and I have performed a sequence of experiments to entangle a single living bacterium with light, as well as another set of experiments to combine a tardigrade (a kind of micro-animal that we will meet again later) with two superconducting qubits. These experiments succeeded in showing that living systems can behave quantum mechanically, though they are still very (very) far away from the states Schrödinger had in mind.

You might have noticed one thing earlier that I was hoping to sneak past you. When I talked about Newton's first law, that a particle moves in a straight line when there are no forces around, I discussed the formula of distance equals speed times time. Then I proceeded to say that speed and distance are q-numbers. But what about time? Heisenberg decided not to quantize time! And it worked as far as atomic physics was concerned. However, this could have been a fluke.

Thinking from our present perspective, this lucky accident could have been for two reasons. One is that the effects of quantizing time might be small and maybe they are not detectable in any experiments we have done or are likely to do in the near future. The second reason is that time might not be a fundamental entity like velocity. Maybe we don't need to quantize time because there is no such thing as time in the first place! Both possibilities are mind-blowing. We will encounter the mysterious nature of time again when we discuss experiments. It is a problem whose solution demands we rely on quantum information tools.

NONSENSE NEVERTHELESS

Einstein was completely perplexed by these developments in quantum physics. When he complained to Heisenberg about some of its seemingly paradoxical features, Heisenberg replied that he was actually following Einstein's intuition to get rid of all underlying features that were not observable. "It was all following your philosophy," said Heisenberg. Apparently, Einstein replied: "It's still nonsense nevertheless!" This extended to the q-reality of quantum physics, because Einstein thought he had a way around the uncertainty principle. He created a thought experiment to illustrate how one can measure the position and velocity of a particle even in quantum physics!

This was the basis of the famous EPR paradox paper, which Einstein wrote with Boris Podolsky and Nathan Rosen. It is interesting that the paper came out in the same year (1935) that Schrödinger's cat paper appeared. Both are underpinned by the phenomenon of entanglement, and the spirit of both is basically "Surely things cannot be that way." Schrödinger changed his mind later and fully embraced the quantum reality, but Einstein never changed his mind.

The EPR paradox was based on a simple argument, some aspects of which we have already discussed. Imagine that two identical particles are perfectly entangled in their positions and velocities. We could, for instance, think of some decay process that results in two identical particles flying in opposite directions with the same speed. Their positions are correlated, too, in the sense that they will always be the same distance away from the point of decay. So we can measure one particle to determine its position and measure the other particle to determine its velocity. However—and this is the crux of the argument—because the particles are entangled, the measured velocity of the second particle must be the same as the velocity of the first particle (in the opposite direction). And if the particles are far apart, surely—say Einstein, Podolsky, and Rosen—the measurement of one cannot affect the other one.

Clever. Einstein, with the help of Podolsky and Rosen, was using quantum mechanics to beat quantum mechanics—using entanglement to beat the uncertainty principle. But, sadly for Einstein, this cancellation of "bad" features didn't work out. The problem with the EPR argument is that there is no violation of the uncertainty principle at all. Although we can indeed measure both position and speed with the two measurements, this does not mean that these values hold simultaneously. In fact, they don't. You can do this even with a single particle: you can measure first one property and then another, but this is not the same as knowing the two properties at the same time. Even more, the order in which you measure things will affect the results. Because we are dealing with q-numbers, first measuring the speed and then measuring the position will give you different numbers than if you first measure the position and then measure the speed.

The punch line of the EPR argument is that quantum physics must therefore be either inconsistent or incomplete. Inconsistent because you can use one of its features (entanglement) to invalidate another one (uncertainty). Incomplete because the EPR argument seemingly (but not really) allows us to know the positions and velocities of particles; however, entanglement apparently does allow us to do so. Hence, it concludes, there must be some hidden variables out there that give us the full picture of reality without any restrictions, just as we have it in classical physics. According to the EPR argument, these hidden variables pertain locally to each particle, and they are c-numbers.

This remained Einstein's vision for the rest of his life. Quantum physics, however, moved in exactly the opposite direction. The EPR experiment did not remain just a thought experiment. We have actually shown that the EPR argument fails in all sorts of systems.

FOR WHOM THE BELL TOLLS

The EPR paper led to a more quantitative prediction by John Bell. Some thirty years later, he derived an inequality known as Bell's inequality. He showed what criteria a theory based on

Einstein's local hidden variables must satisfy. Remember that the word *local* means that the hidden variables, which are just c-numbers in the classical description, actually characterize the system locally in the place where the system is situated. This is the key to all the fuss regarding entanglement. We don't want the information about an atom on Earth to exist within another atom on Mars. It's as simple as that.

The Bell premise is as follows: if you make such-and-such measurements on one particle and so-and-so measurements on another particle, assuming that the measurements cannot affect each other because no light can travel between the particles, then the outcomes of these measurements (for simplicity, assumed to always give us the values +1 or −1) cannot, when put together, exceed 2. The reason is that when one particle yields the value of +1, so could the other, while when the first yields the value of −1, so could the second one. Together, we have $(1 \times 1) + (-1 \times -1) = 2$.

The Nobel Prize in physics in 2022 was given to three physicists, Alain Aspect, John Clauser, and Anton Zeilinger, "for experiments with entangled photons, establishing the violation of Bell inequalities and pioneering quantum information science." But it is not just photons that have violated Bell's inequality and therefore proved EPR wrong and quantum physics right. We have violated it with atoms, subatomic particles, molecules, and so on. More or less anything that we physicists have been able to entangle can be shown to violate Bell's inequality. These various quantum entangled particles have been tested, and they reach a larger number, $2 \times \sqrt{2}$.

What is the conclusion of all this? It is that we have to give up either of the two assumptions leading to Bell's inequality. Either each local system in the experiment cannot be independently described by c-numbers, or there is something traveling faster than the speed of light telling one system how to adjust to the measurements performed on the other, faraway system. The latter basically means that relativity goes out the window. So either c-number reality or relativity (or both?) has to be abandoned.

My hunch, and the hunch of most other physicists, is that it's not relativity that fails. So the only conclusion left is that the underlying elements of reality cannot be c-numbers. Rather, they have to be q-numbers. That's the only reasonable way to explain the violation of Bell's inequality and not end up violating relativity. And this is very important, as it instructs us how to proceed to test for new physics.

There are other ways to account for the violation of Bell's inequality, even if one keeps the c-numbers and relativity. However, they seem to me completely unreasonable, involving assumptions that are too farfetched to be taken seriously. These are all known as loopholes because they still allow for a "classical" account of the violation of Bell's inequality, although, as I say, not many researchers take them seriously.

One is retrocausality, which is the notion that an effect could precede its cause. The TV goes off before you've pressed the remote control. Or you die, only then to be born later. I would love to tell you a joke about retrocausality, but I won't, because you didn't like it. Seriously, though, I want to discuss this and a number of other (what I would consider to be) dead ends that have been explored in physics (and are still being explored), especially in relation to quantum physics.

Retrocausality captures the notion that there are things in the future that can affect things we experience in the present. It is clearly something for which we lack any hard evidence, though there is a great deal of reported anecdotal evidence out there (I find none really believable, but it's up to you to make up your mind). Unsurprisingly, then, retrocausality is not independent of other quirky notions such as time travel, or pseudoscientific claims linked to clairvoyance, extrasensory perception, premonition, and so on.

Retrocausality could certainly come in handy when trying to understand entanglement. The issue is—in very simple terms—whether we can avoid entanglement and still explain the results of measurements (which all confirm entanglement so far) by using states that are not entangled. Could we have a local description of the state of one system and another local description of the other system in terms of c-numbers? Bell's inequality tells us that c-numbers won't work unless there is something faster than light going between the entangled particles. This includes signals from the future, which are—under some conditions—equivalent to superluminal communications. How does this work? Well, we take the measurements on the particles (one being that of the positions and the other being that of their speeds) as the initial (past) values of these quantities for the particles themselves, but also as the coming-back-from-the-future values for both particles (in order to capture what entanglement between them would accomplish).

This way, each particle is characterized by two c-numbers that pertain locally (in space) to the particle. However, they are highly non-local in *time* given that one of the two numbers (for each particle) has to be communicated back from the future. It would be an interesting suggestion if it weren't for the fact that no one has ever received any signals from the future. It also doesn't do much for understanding entanglement, as it requires rewriting history to account for what has already happened—that is, once we have done all our measurements. This defeats the point of science, which is all about predicting the future. The past doesn't have an issue with indeterminacy: if you have already made your measurements, you know the outcomes you got. They are recorded in your memory, your log book, or anything else that generates a track record.

Besides retrocausality, the only other way out of the q-number reality is to endorse superdeterminism (a phrase I don't like, as I think determinism is already super in itself). Superdeterminism simply says that all results are encoded in the initial conditions of the universe. You think you are able to measure arbitrary things in the future (a position of this, a speed of that, etc.), but you don't actually have the freedom to do so since it is all predetermined. Superdeterminism is therefore another loophole.

Like retrocausality, superdeterminism doesn't really do anything for science, since it lacks any predictive power. Answering the question "Why did this happen?" with either "Because this is what the initial condition told us" or "Because this is what the future outcomes told us" is not a good explanation. And neither can tell us in practice what will happen in advance of the measurements we make. Despite their critics, I think entanglement and q-numbers will still be with us in 2026, and we will learn that an even greater range of phenomena are genuinely quantum, just as our current understanding suggests. I don't dare predict (at least not just yet) that experiments will confirm in 2026 that gravity is also quantum. That's probably too soon. But—more importantly—I've had absolutely no communications from the future telling me otherwise.

One concept that *did* do a great deal for science was that of the field, which was invented in physics in order to preserve the idea of locality. Fields came about as an answer to Newton's (and many other people's) worries about how gravity seems to act at a distance, since massive objects attract one another across the vastness of space without any (apparent) mediation in between. In order to remedy this, Faraday came up with the idea of the field: an object affects the field only in its immediate vicinity, and this disturbance then propagates through the field much as a wave would when a stone is thrown into a pond. This wave of disturbance in the field ultimately reaches another object, and the field disturbance then affects it locally, at the point where the other object is (as a lotus leaf in the pond would be affected after the wave caused by the stone reaches it). That way, any action at a distance is avoided.

Exactly the same is true for quantum fields, the only difference being that what does the waving in the quantum case are the q-numbers pertaining to the field.* Here the speed of propagation is of course the speed of light. But the speed itself is irrelevant for the concept of locality. Instead, we simply need to require that acting on one subsystem (whatever this is, be it an atom, a photon, or a region of space) cannot instantaneously affect another subsystem. We will spell out below exactly what this means, but suffice it to say that this principle of Einstein's locality is also true in non-relativistic quantum mechanics. If we have two qubits—and remember, a qubit is the quantum analogue of a bit, and so it is the fundamental unit of quantum information and quantum computation—located at two different places, then acting on one of them cannot change any

---

* In the case of the electromagnetic field, these would be the operators characterizing the vector potential, the electric field, and the magnetic field.

property (any q-number) of the other qubit. Nevertheless, it is sometimes mistakenly said that quantum physics violates Einstein's locality. Quantum physics violates an altogether *different* idea of locality, called Bell locality. In Bell locality, things are described by c-numbers (not q-numbers) locally. And, of course, we know that the violation of Bell's inequality is in direct conflict with the existence of the c-number-based local reality.

The fact that quantum physics is Bell non-local and general relativity is not poses no obstacle to their unification.[*] When the fundamental entities in general relativity are quantized, the resulting theory of quantum gravity also becomes Bell non-local. Any classical theory that is quantized is automatically—by its very construction—Bell non-local. Failing to appreciate this has been the source of much trouble in the foundations of quantum mechanics. We will discuss this at length when we talk about how to test the quantum nature of the gravitational field.

THE BEST SO FAR

Our best current understanding of the universe comes from quantum field theory, without which we could not account for many of our experiments.

The seventeenth-century Dutch physicist Huygens was once ill and lying in bed when he noticed that two of his clocks (they only had grandfather clocks back then), positioned on the same wall opposite his bed, started to oscillate in sync. No matter how out of sync they started, they would always end up oscillating in phase. Ever a man of science, Huygens would

---

[*] No more than it is an obstacle to quantum electrodynamics, that is.

experiment by putting them on separate walls and starting the pendulums randomly, after which he would observe that they no longer ended up in sync. The only possible conclusion was that each clock produced vibrations that were transmitted by the wall and then to the other clock. This was how they synchronized. When they were far apart, this no longer worked, because the vibrations died down before reaching the other clock.

Something similar happens when you have two tuning forks close to each other (have a look at YouTube). You hit one fork and see that, shortly afterward, the other one starts to vibrate, too. This is true only if they are of the same frequency. Otherwise, nothing happens. In this case, it is the air in between that picks up the vibrations and transmits them— at the speed of sound—to the other fork.

This simple concept forms the basis of quantum field theory. Quantum field theory is all about harmonic oscillators like the grandfather clocks, only quantum! The universe is made up of a stupendously large number of quantum harmonic oscillators, each representing a degree of freedom of the respective underlying field (the electromagnetic field, the electron field, and so on).

Take an atom absorbing (or emitting) a photon. An electron inside an atom is a harmonic oscillator. A photon is an excitation of the electromagnetic field, which is another harmonic oscillator. These two oscillators couple to each other through what is known as the dipole-electric field coupling. The details of this don't really matter and you can look it all up if you really want to know, but the crux of the matter is that it's the same situation as with the two tuning forks. In fact, even the underlying mathematics is the same in both cases; it's just that the tuning fork can be understood with c-numbers and the atom and the field require q-numbers.

When the frequency of the photon is not the same as the frequency of the electron, the two are not in tune, and so the atom cannot be excited by the photon (the photon is "off resonance"). When we excite an atom, it emits a photon that can then be absorbed by another distant atom if they are of the same frequency (or "on resonance"). The atoms are also like the two grandfather clocks, with the in-between electromagnetic field playing the role of the wall that allows the energy (photons) to propagate in between (granted, photons go at the speed of light, not sound). Huygens would have understood the quantum physics of the light-atom interaction immediately.

All interactions in quantum physics are described this way. There are particles (pendulums) that interact with other particles (other pendulums) through mediators (which—surprise, surprise—are chains of yet more pendulums).

In classical field theory, the electric field values and the magnetic field values are defined at each point in space by three electric field numbers and three magnetic field numbers, all of which change in time. That is the electromagnetic field. When this is quantized, the electric field and the magnetic field become "operators" that generate q-numbers. And it is these operators that exist at every point in space and change with time. That is the *quantum* electromagnetic field.

Although quantum fields are very different from classical fields, they do not allow action at a distance. Quantum fields have quanta (photons in the case of the electromagnetic field and gravitons in the case of gravity, though the latter have not yet been observed). For the quantum electromagnetic field, the electric field and the magnetic field cannot be

76

measured or specified simultaneously, just as the position and momentum of a particle cannot be measured simultaneously in standard quantum mechanics. One of the counterintuitive consequences of the fact that the fundamental entities in the universe are quantum fields (which are really just our old friend, q-waves) is that particles should not be thought of as being elements of reality. In the same way that the position and momentum, as c-numbers, cannot be elements of reality in quantum physics, the same goes for the number of particles that constitute a given system. In other words, the number of particles itself ought to be a q-number, rather than some definite, single-valued number.

For instance, we can have one frequency of the quantum electromagnetic field in the state where we have a superposition of one and ten photons. This means that the frequency is excited by one and ten units of energy simultaneously. We can even have a superposition of no photons (the so-called vacuum state) and one photon. Classically, these examples are clearly impossible. More than that, if we have superposed states in which each branch has a unique number of photons, then surely photons themselves cannot be fundamental. The same goes for all other particles, which can be thought of as just energy excitations of their respective quantum fields.

This is mind-blowing stuff, that particles are not elements of reality, but what happens when we make a measurement of one of those states that is in a superposition of no particles and one particle? The above discussion of measurements being entanglements between two quantum systems suggests that there will be one state in which we will not see any particles, but at the same time there will be another state in which we will see one particle. And these two different branches, each containing one of the two possible outcomes, themselves exist in a superposition. In each branch, the existence or nonexistence of the particle is a "classical fact" with respect to the system that did the measurement through entanglement.

In other words, if we insist on the view that particles are definitely there or not there (and not simultaneously both), we would not be able to reproduce all the experimental results in quantum mechanics. There is even a Bell inequality that arises out of the insistence that particles are either there or not there, and we have already had experimental violations of these inequalities with photons. Björn Hessmo, Pavel Usachev, Hoshang Heydari, and Gunnar Björk from the Royal Institute of Technology in Sweden did this way back in 2004.

Fundamentally, it is straightforward to see why special relativity forces us to allow superpositions of particle numbers. Relativity tells us that particles are not conserved. What is conserved is the total energy. So an electron and a positron can actually annihilate each other in order to produce two photons whose total energy is the same as that of the initial electron and positron. We must therefore allow for a superposition of the state in which the electron and the positron exist but photons do not, and the state in which the two have annihilated each other and produced two photons.

Even more, antimatter particles such as the positron—whose properties are all the same as those of the electron (mass and spin) but which has the opposite charge of the electron—must also exist, according to relativity. This is because the electron q-wave, taken on its own, would be able to propagate faster than the speed of light. If an electron is initially confined to a finite region of space, then there is a non-zero probability that its wave is found at the next instance at an arbitrarily faraway place. This would seemingly imply an arbitrarily high speed of propagation, thereby violating relativity. Spooky action at a distance indeed. However, the electron q-wave must also contain a component of the positron; it is the superposition of the two, the electron and the positron, that constitutes the correct q-wave.*

REAL BUT NOT TRUE

By a funny coincidence, as I was completing this chapter, I came across a beautiful little book in my local secondhand bookstore. It was written by Charles Darwin in 1931. You read that right—and, no, I haven't made a chronological typo. This is not the Charles Darwin of evolutionary fame but his grandson, who himself was a distinguished physicist. The book, titled *The New Conceptions of Matter*, is by far the clearest popular exposition of quantum physics I have ever read.

The book presents a fantastic account of the view that all matter is made up of waves. It seems amazing to me that at the time Darwin wrote it, quantum physics was only five years old. During that short period, Darwin not only absorbed all the new, revolutionary ideas of quantum physics but also made novel contributions to it and gave us the germ of what I believe to be the best interpretation of quantum physics ever.

Darwin thought of particles as merely a useful fiction. He gave the following analogy. When a string on the piano oscillates 256 times per second, we perceive this as the note middle C. We don't hear a wave vibrating rapidly (whatever that might mean) but instead have a sensation that feels singular and

---

* It is this antimatter component that cancels the material components traveling faster than the speed of light and restores relativity within the q-waves.

definitive—much like a particle. Likewise, when the vacuum vibration of about 20,000 waves per centimeter hits our eyes, we don't perceive a rapid modulation of a propagating wave of light; instead, our brain interprets this particular vibration of the electromagnetic field as simply the color yellow.

The note C and the color yellow are our subjective experiences of reality. They are convenient fictions, shortcuts that our brain circuitry invented to make us into highly functional animals. They seem real to us, but they are not really true (a phrase I unashamedly lifted from Buddhism). This is so for all our perceptions. When we put our fingers close to fire, we feel hot. We don't feel that the atoms of the air are more rapidly hitting the atoms of our skin, which is really what temperature is all about.

Darwin says that particles are still useful fictions to sometimes think in terms of, but one must always remember the underlying wave nature of everything. Otherwise, it would be impossible to explain the host of phenomena that matter clearly exhibits, such as quantum tunneling (the ability of quantum objects to penetrate barriers that classical physics tells us they don't have enough energy to penetrate), matter wave interference (the fact that electrons and atoms behave like photons), stability of matter (the fact that solids can exist), its bulk properties (such as heat and electrical conductivity), and so on. The objective reality of an electron is that it is a wave, while we subjectively perceive it as a particle. Likewise, the electron wavelength is its objective property, while the momentum is our subjective convenient fiction (momentum being something that particles possess). The electron wave frequency is another of its objective properties, and this corresponds to its energy, which is a useful fiction if we think of the electron as a particle. And so on.

Of course, the question of why we perceive the outside world the way we do goes beyond the realms of physics and will probably only be fully answered by chemistry, biology, neuroscience, and computer science all working together with physics. But for the purposes of our discussion, we can think of it as an evolutionary accident, not a necessity of the laws of physics (another example of an evolutionary accident is that humans' strongest sense is sight, but for bats, it's their hearing apparatus).

As before, the conclusion is that there is no need to think of ourselves as special observers, as something outside of physics. We are an integral part of it, made up of q-numbers like anything else. When we perceive something, it's really just one q-wave interacting with another. This is all there is to making measurements in quantum physics. No problem at all.

THE KEY TOOLS OF QUANTUM INFORMATION

At the most basic level, we need two completely distinguishable states (on or off, 0 or 1) in order to talk about a single bit of information. This will allow me to send you a message that communicates either a yes or a no. Shannon showed that, with enough bits, anything can be communicated, and he calculated the rate of doing so faithfully under different circumstances. It turns out that classical bits are a special case of the more general quantum bit (qubit), for which there are also two perfectly distinguishable states; however, one can make any arbitrary superposition of them, too.

Understanding the information content of our theories in physics is crucial. And I don't just mean this in terms of their information-processing powers. Yes, there is that, too, in the sense that quantum information processing is potentially exponentially more efficient than the classical version. This means that there are tasks—some of which have relevant applications in, say, cryptosecurity, chemistry, the pharmaceutical industry, and material science—that can be performed much faster on a quantum computer. I talked at length about such things in my book *Decoding Reality* about fifteen years ago.

Since then, quantum technologies have developed rapidly, and major companies like Google, IBM, and Microsoft, as well as countless start-ups, are now racing to make the first universal quantum computer. When I wrote *Decoding Reality*, I was confident that quantum computers would be possible. Now I think I've been vindicated. It's really just a matter of time. The problem is no longer whether quantum computers are allowed to exist as far as physics is concerned. Now it is simply a question of engineering.* But here I want to focus on the physics of it all.

A MAN WITH A RIFLE

Within a physical theory, which elements of reality are capable of carrying information? We can say that something has information if the states encoding this information can be copied—faithfully, without errors—into something else. In quantum physics, things we call observables, otherwise known as our friends the q-numbers, have this property. Different values of a given observable are fully distinguishable from one another, and this is what makes them able to be copied. As I mentioned, each observable is represented by a q-number, and

---

* Photons, atoms, and superconductors are all used to make qubits.

these q-numbers can be copied from one quantum system to another. A perfect example of copying is the quantum measurement we discussed above: it is simply the copying of the information within one q-number of the system into another q-number belonging to the measurer.

In classical physics, all c-numbers can be simultaneously copied from one system to another. The position and velocity of one system can be imprinted onto another system. This is what a police camera does when it catches you speeding: it measures the position and velocity of your car and copies the values into its report, which then gets copied into the court statement, and so on. In quantum physics, the uncertainty principle tells us that we can copy the position *or* we can copy the velocity, but we cannot copy both. This is just another way of saying that the elements of reality in quantum physics are q-numbers. My colleagues Chiara Marletto and David Deutsch call quantum systems "superinformation media."

What distinguishes superinformation from regular, classical information? Every superinformation medium consists of many q-numbers. Each of these q-numbers is an information medium in its own right. However, it's not always the case that when you bring together systems that are individually described by q-numbers, the combined system is an information medium! In other words, combining two things that can be copied does not always result in a new thing that can be copied. This is the information-theoretic way of understanding the uncertainty principle of quantum mechanics: you can measure a system's position (something that can be copied) or its velocity (another thing that can be copied), but you can't measure them both simultaneously to arbitrary accuracy (the things that can be copied individually cannot be copied together).

The information-theoretic view is also the best way of understanding quantum measurement. Because we can copy the position of a particle into another particle, but we cannot at the same time copy information about its velocity, the position and velocity states are not fully distinguishable. Each (the position and the velocity) is an information medium in its own right, but when taken together they become superinformation. The totality of q-numbers is superinformation.

The second key aspect of quantum information is what might be called coherently controlled information processing. Namely, if we have one way of doing something, and another way of doing the same thing, then, by using a single qubit, we can make a superposition of the two processes (by definition, qubits have two distinguishable states that can be superposed). This simple quantum information trick is missed by most practitioners of quantum physics, yet it makes a huge difference as to what we conclude is doable. We frequently imagine one observer seeing one thing and another seeing something else, but we seldom contemplate that the two could be in a superposition, existing at the same time. I am convinced that not being aware of the fact that we can superpose any two processes is at the root of why some people think that there is still the measurement problem in quantum mechanics. It also prevents us from properly understanding the nature of identical particles—a group that includes the fermions (particles like electrons, protons, and neutrons)—and ultimately how gravity ought to be quantized.

The third important aspect is that information, whether classical or quantum, obeys the principle of local tomography. Here, *tomography* is just a fancy word for estimating the state of a system, and *local* means that it is done by measuring a system's constituent parts at the place where the system is. At the heart of local tomography is the fact that q-numbers pertain to local systems, and when we measure q-numbers of one system, we do not change the q-numbers of another system. The same is true in classical physics but with c-numbers. When we measure the position of my car, we do not change the speed of yours. Likewise for all other properties. In the classical world, local tomography works like this: if I tell you everything about my grandmother and then I tell you everything about my grandfather, I've basically given you all the information about my grandparents. The surprising thing is that, despite entanglement, the same is true in quantum physics: we can infer the state of a quantum system composed of many subsystems by locally measuring each subsystem and then gathering all the data together. The principle of local tomography says that it suffices to measure the particles separately to understand their joint state of affairs. All the experiments regarding the violation of Bell inequalities that have been performed so far, including the three that resulted in a Nobel Prize in 2022, have been performed effectively using local tomography.

The fourth and final principle we need is that information is always conserved. It cannot be erased and it cannot be created. In classical physics, c-number-based information is conserved, while in quantum physics it is the q-number-based information—the information residing in superinformation media—that is conserved. In that sense, information joins energy as a quantity in physics that cannot be altered in any interaction. This is true only in totality, and information can of course be gained by a measurement in a relative way.

However—and this is the biggest *however* in this book— when a system obeying one kind of information is combined with another system obeying a different kind of information, then the principle of information conservation seems to no longer hold. This is why any description of the world that involves two theories with different kinds of information can never work if the principle of information conservation holds globally. This is ultimately because when we have two theories with different information capacities, the contents of one theory cannot be faithfully copied into the other. In order for information to be preserved, the theory with a higher information content must always win out (in this case, superinformation wins over classical information). The only way to describe faithfully two media of different information capacities is to assume that the medium with the lower capacity actually does possess the same higher information capacity, though this fact may be concealed. In other words, superinformation is more fundamental than classical information. This is a major reason we think gravity will need to be quantized, as classical theories of gravity do not allow for superinformation at all—if classical information is really just quantum information in disguise, then information is conserved after all.

This reminds me of a line in one of my favorite Westerns, *A Fistful of Dollars*, in which Clint Eastwood's character says: "When a man with a .45 meets a man with a rifle . . . the man with a pistol is a dead man." Here the man with a pistol is general relativity. We will see this principle in operation when we discuss ways of testing quantum gravity.

Incidentally, conservation of information also contradicts the Copenhagen interpretation of quantum physics (as well as various collapse modifications in which quantum physics, under some conditions, is forced to become classical). According to the Copenhagen interpretation, the experimental physicist who probes a quantum system determines how that system will ultimately behave via his choice of which observable to measure. Will it be forced to have a definitive position? Or will it be measured to have a definitive velocity? In other words, the observer is classical and imposes the external classical context on the quantum system. This is at the heart of the Copenhagen logic. As I said, this contradicts the principle of information conservation because quantum systems can handle more information than classical ones. And the implication is that, under some circumstances, the information capacity of quantum systems should be lower than what quantum information theory says they have. Copenhagen logic implies that a quantum system can only consistently stay fully quantum if it never couples to any classical systems.

Why did the Copenhagen view become so dominant? Partly it is because of Bohr's charismatic personality and the way he approached promoting his own views. But it is mainly because the Copenhagen interpretation is pragmatic, doesn't ask uncomfortable questions, and gives us the simplest prescription for how to calculate the right result in many textbook-like examples of quantum phenomena. This is why Dirac said, in his usual dry manner: "Bohr's doctrine may be useful for students preparing examinations, but not for physicists doing physics."

THE QUANTUM UNIVERSE

Before I show you one example illustrating the power of these tools, I will take a short detour to talk about the role and origin of principles in physics. This will provide the bigger picture to help us understand how the quantum tools, combined with the properties of quantum systems, imply that the whole universe has to be quantum. If the principles of quantum information and q-numbers are truly more fundamental than the principles that govern classical information and c-numbers, then other principles we hold dear should be consistent with this q-number view of reality. Hold that thought.

What's remarkable about principles like those of local tomography or information conservation in physics is that they are like axioms in geometry. We can't say why they are true; they just seem to be. And we rely on them heavily. As I outlined earlier, Einstein derived both special and general relativity from a handful of principles. The principle of undetectability of uniform motion and the supremacy of the speed of light gave us special relativity, while the principle of general covariance and the equivalence principle gave us general relativity. Physics is full of such principles, and they often allow us to reach elegant conclusions and deeper insights than mere laws of motion ever could.

But it wasn't physicists who invented reasoning from principles—it was actually the ancient Greeks.

Take the reductio ad absurdum. It's a method of proving that something's got to be the case by supposing the opposite and showing that this leads to a contradiction or an absurdity (hence the name). There is a passage in Plato's writing in which he wants to prove that in the world of forms, there is only one of each perfect thing. There are many cats in our world, but there is only one perfect form of a cat in Plato's world of ideas. All real-world cats are just imperfect copies of the ideal cat. And so on, Plato would say, with everything else we encounter (I am oversimplifying Plato a smidgeon just to get the point across).

How did Plato prove that there was only one ideal cat? By supposing that this was not true. Start by assuming that there are two such cats. But then surely, says Plato, they would both have to be made in resemblance to yet another more perfect cat. Hence a contradiction, and so there can be only one such perfect cat. While Plato wasn't the inventor of this trick, he used it a lot in his own dialogues and brought it to the masses.

They say that it was the Renaissance that unshackled the human spirit from the hold of medieval Christianity (*renaissance* means "rebirth" or "reawakening"). But that reawakening probably happened when it did because of the fall of the Byzantine Empire, which led to ancient Greek texts that had been preserved in the Middle East flooding into Europe. It was then that the medieval Italians (there was no country called Italy then, to be sure) realized how ignorant the Middle Ages had made them in comparison to the ancient Greeks. It was their frantic consumption of these newly available ancient Greek texts that set their spirit free, prompting them to start the Renaissance and lead the rest of us into modernity.

I must admit that I envy Plato and that ancient bunch. They had the fantastic Mediterranean climate, they enjoyed tasty food and wine, and they were great sailors, warriors, and statesmen. But above all, they would discuss important philosophical issues out in the agora while eating and drinking copiously (they called these gatherings symposia, which derives from a verb that literally means "drinking together," and I strongly feel that our modern-day scientific symposia are rather poor copies of what the old Greeks used to do).

But there is one ancient Greek principle above all, I think, that set us on the path of science. It is that every phenomenon needs a reason. Nothing can be just so! It is not enough to note that the sky is blue and the Earth is a sphere. We need to explain why this is so. And explain we can, for both have good reasons for being the way they are. The ancient Greeks were confident that everything has to be explained and *can* be explained purely by reason (some two millennia later, Gottfried Wilhelm Leibniz lifted this idea from the old Greeks, refined it a bit, and called it the "principle of sufficient reason").

I recently came across an argument by one of the pre-Socratic philosophers (I think it was Thales) that, to my surprise, effectively amounted to using symmetry in order to argue for conservation.

"What?" I hear you say. "Symmetries imply conservation?" Yes, we know this well in physics, but we attribute the result to an early twentieth-century German mathematician called Emmy Noether. She proved that if the laws of physics have certain symmetries (in other words, if they remain the same when we change some things in the equations), then momentum, angular momentum, energy, and so on are actually conserved (meaning that they always have the same value no matter what else might happen). This is a surprising and profound result. It underpins much of modern physics and the way we think about it.

The ancient Greeks didn't have the mathematical sophistication of Noether (although they were not bad, either), nor did they understand physics the way an early twentieth-century physicist did (if anything, their physics was worse than their mathematics). However, the germ of the symmetry-implies-conservation idea was already there. They gave us the key concept.

For instance, Thales wanted to explain why the Earth is stationary and at the center of the universe (yeah, I know; I did warn you about their physics). It was not enough for him to observe that the Earth seems to be stationary and at the center;

he wanted to explain why this is so. His answer: because the universe is the same in every direction! After all, why would the Earth prefer to go in one direction rather than another, since they are all the same? It is for this reason (the directional symmetry of the universe) that the Earth stays where it is. This logic is so beautiful that all the misconceptions regarding the position and motion of the Earth are immediately forgiven.

If you've understood this idea, you are ready to understand what we call the principle of the conservation of momentum. A particle moving in empty space continues to move at the same speed (momentum is just the product of its mass and its speed). Why? Because empty space is the same everywhere! If a particle stopped, slowed down, or sped up at some point, then this point would be special compared to all other points. But there is no such special point if the universe is everywhere the same. Ergo, momentum is conserved.

You might be thinking that this momentum argument is due to Thales, too. However, it is due to St. Thomas Aquinas. Our best *modern* argument for momentum conservation is due to Noether. The fact that space is the same everywhere is called translational symmetry (the laws of physics here would be the same if we were to displace everything by any amount of distance). Momentum conservation is a consequence of this.

The laws of physics are presumably the same now as they will be in a minute or an hour. If you assume this principle— that the universe remains the same if everything is displaced in time—you will conclude that the overall energy is conserved. Amazing! Energy conservation follows from the fact that no instant in time is different from any other instant in time (if energy increased at a particular time, then that instant would be special, which contradicts our assumption that all times are the same—notice Plato's reductio ad absurdum all over the place). Thales didn't know that energy is conserved (for starters, he didn't even know what energy was), but he would get the logic in an instant if we could bring him back from the dead (I wish).

Yes, we definitely know more than Thales and his philosopher buddies did. However, something tells me that if Thales were back, after patiently listening to all our evidence for quantum physics, he would still almost certainly ask us, "But why is the ultimate reality quantum mechanical?"

We would have to admit to Thales that we simply don't know. In other words, we don't have a simple principle from which to argue that the world must be quantum the way that Thales argued for the central position of the motionless Earth. Is there such a principle? Many believe that there is, and some of us, including me, believe that it is information-theoretic in nature. But we can make a further damning argument against the principles of the Copenhagen interpretation and for a principle that the universe is totally quantum. And to do that, I must tell you about how energy and momentum conservation work in quantum physics.

It won't surprise you to hear that both are conserved in quantum physics, just as they are in classical physics, although of course they are q-numbers rather than c-numbers.

But remember how it is possible to superpose different numbers of particles? That seems to immediately cause a problem for my claim. Suppose we make a state of light that is a superposition of zero and ten photons. The average energy is five photons' worth. But when we measure this state to check the number of photons, there is a probability that we obtain ten photons, which is twice as much energy as we started with!

Of course, there is also a probability that we get zero photons, which is less energy than we started with. This too seems to violate energy conservation. Only on average, it would seem, is energy conserved.

This would indeed be a weird and unsatisfactory state of affairs if it were true. And it quickly reveals the flaw in thinking that we, or our devices, are not quantum mechanical. As a result of that kind of thinking, we didn't include the measurement apparatus in our considerations. We only looked at the energy contained in the state of light; we didn't include the energy required to measure it. When ten photons are detected, the apparatus absorbs them at the expense of the electromagnetic field losing them (remember, photons are excitations of the electromagnetic field). If you follow this logic and take a quantum apparatus into consideration (if you take the universality of q-numbers seriously), you will actually realize that energy is conserved exactly in quantum physics—and not just on average.

There is a great deal of debate about issues such as this one, but all the apparent problems (like a potential non-conservation of energy) arise only when we treat half of the things quantum mechanically and the other half classically. In other words, all of the apparent problems disappear if we think of the whole universe as being quantum.

Unfortunately, the principle of quantum totality still wouldn't answer Thales's imagined question: "Why quantum?" As I said, we don't know, but we can always speculate. After all, the ancients dared to speculate, and that's what ultimately empowered us. What would Thales have said about quantum physics? My feeling is that he would say something like this: if you have two situations that you cannot tell apart, such as a photon arriving at us from one slit or another, then you have to acknowledge that both could be true at the same time. It is this democratic principle (which would definitely appeal to the Greeks, who themselves invented democracy) that is at the core of the superposition principle. If there is no reason to have zero photons rather than ten photons, then probably we should have a superposition of the two. That might be the most important principle of quantum physics: anything that's not prohibited should ultimately be allowed to exist.

———————

We've journeyed through some of the deepest theories in physics, many of our most cherished principles, the problems that we physicists currently need to solve, and the difference between q-numbers and c-numbers (and some of the differences between quantum information and classical information). The hard work is done and the stage is set. We are now ready to delve into the experiments that could serve as portals into a new reality. And what a ride this will be! Keep your rifles loaded and your powder dry: the first experiment will shoot down nothing less than the measurement problem.

Chapter 4

THE "EVERYTHING IS A Q-WAVE" INTERPRETATION OF QUANTUM PHYSICS

A Nightmare of an Eminent Person: A Supporter of the Copenhagen Interpretation Goes to Hell

This is a story about a dear friend of mine, Professor Gustav Klimtovich, who is a world-leading researcher in theoretical quantum physics. Gustav is based at an old Central European university that recently underwent a renaissance (climbing up two hundred places on the Times Higher Education university list), in no small part due to Gustav's groundbreaking work in the foundations of quantum mechanics. To my great surprise, however, even an illustrious person like Gustav has nightmares. He told me about a recent one involving Hell and the Devil. Gustav's story made such a strong impression on me that I decided to communicate it to you. I will tell it to you as he told it to me: in the first person (on his behalf).

I fell asleep one night and dreamed that I died. Previously, I had thought that it was impossible to dream about one's own death. Anyhow, despite the fact that I am a rather pious, church-going individual, I was being sent straight to Hell. I attributed this unfortunate occurrence to the fact that I had rejected too many good research papers from my competitors, which (one could indeed convincingly argue) might have been morally questionable. I accepted my fate without resistance and didn't worry much about going to Hell. *How bad could it be?* I thought. Little did I know . . .

At the gates of Hell, I was greeted by a funny-looking fella. Goatee, sharp pointy ears (almost invisible, as though glued to his skull), squinting gray eyes, and, above all, really well-dressed (clearly not a physicist, I surmised). I immediately assumed he was some kind of a porter, one of those you encounter when you go to an Oxford college.

He asked me something unexpected. If I was to be sent through a beam splitter, the fella said, which side of the beam splitter would I come out of? *He is definitely a funny guy,* I thought. I said that if it was a fifty-fifty beam splitter, then I suppose that I would have an equal chance of coming out of either end. I told him that it is just a direct application of the Born rule (for some reason, I assumed that this guy would be familiar with the third postulate of quantum physics).

"All right," he said. "Let me escort you to meet His Majesty the Devil." This really was stranger still. I was asked a simple question, and upon answering it (correctly, I assumed), I immediately got to see the Big Guy. I wished real life were like that! Imagine answering one question from your research funding agency correctly and immediately getting an interview with its chief executive. That would be nice, I mused. But then came a surprise.

He opened a door and we entered a room that looked like a standard atomic physics laboratory (an optical table with lots of lasers, beam splitters, some fancy cryogenics—you know the kind of stuff). I always thought that experimental physics was the work of the Devil, but I never expected to encounter firsthand evidence of this fact. Now the fella says, "If I give you two million qubits all prepared in an equal superposition of 0 and 1 logical states, and if you measured them in that logical basis, what would you expect to see?"

Is this guy for real? I wondered. He'd asked me a similar question a while back, so why did he expect a different answer now? Despite being confused about his intentions, I nevertheless proceeded to answer: you'd expect to see one million in the state 0 and one million in the state 1, give or take a few thousand (I assumed that he was familiar with standard deviations, too).

"OK," says he. "The beauty of this room is that we don't have to imagine. What you see in front of us is a quantum computer with one billion qubits (and, no, we didn't get it from Google: it is bona fide made-in-Hell). I will now prepare two million qubits in said superposition and leave the room. You can then measure them one by one and see what happens."

I was game. I must admit, I was very excited to see a fully functional large-scale quantum computer. My God, it *is* possible to make one after all (I had always thought that some kind of decoherence or Penrose-style induced collapse would prevent it). It was easy for me to program the computer to execute the measurements. To my great surprise, all qubits came out with the value 0.

The fella came back and must have seen that I was in shock. "So?" he said. I told him they had all come up 0, which is statistically exceedingly unlikely. The fella was kind enough to let me repeat the experiment again. And again. And again. But no matter how many times I did it, all the qubits always came up in the logical state 0.

How can this be? I wondered. Finally I said to the fella, "Could it be that we are using a quantum computer to show that quantum physics is wrong?" (I remember reading a *Nature* paper by Dr. Henrietta Grünberg saying something to that effect.) "No," said he categorically. "It is your Copenhagen frame of mind that prevents you from seeing the truth."

What? How did he know I was a Copenhagen supporter? He must have read my papers. Or maybe he attended my lecture course on the foundations of quantum physics. *No, wait,* I thought, *he does look familiar after all.* Maybe I had seen him at some conference or other. As I was desperately trying to figure all this out, the fella said: "You need to work with the hypothesis that all of Hell is quantum. You and this quantum computer are just part of an even bigger quantum computer (run ultimately by the Devil)."

"What? All of Hell is quantum?" I gasped. "The Devil is a many-worlder, programming the dynamics of Hell?" He (I always assumed the Devil was a he, but, honestly, I am not that sure anymore) actually let me make all these measurements and then run the Grover amplitude amplification algorithm on the whole lab to amplify the branch with all the zeroes to a 100 percent certainty. I was simply flabbergasted.

But much more than the Devil's experimental prowess, what spooked me the most was that I had been wrong all along regarding the interpretation of quantum physics. If the Devil was able to pull this off, Copenhagen cannot be right! I kept deterministically landing in the state that was exponentially unlikely according to Max Born, but instead of refuting quantum physics, this actually proved its validity!

I was desperate, and cried out, "Bohr, why hast thou forsaken me?" At that point, just as I uttered those words, everything around me came to a halt. Then it all vanished in a puff of smoke. There simply was nothing left. Not even the quantum vacuum.

I woke up. Luckily, it took me only a second to realize what had happened. I was right to be a Copenhagen supporter. There is, after all, a classical frontier. It is defined by God. He (again, God surely is a he, too) must have made me collapse into a single universe that makes sense and where I am not having nightmares about Hell, the Devil, and many worlds. Phew. I always knew there must be the ultimate classical frontier. The universe cannot be in a quantum superposition, since it is, thankfully, always observed by God. God provides the necessary collapse needed to get rid of all the devil-ish quantum entanglements.

I was relieved. And, as I always do when I wake up, I turned on my radio. It was ten o'clock—news time. The voice said, "Today, Google has announced its first one-billion-qubit quantum computer." *Damn it*, I thought. *I knew I should have bought more of their shares a couple of years back.*

OBSERVING THE OBSERVER

Here I would like to present the first kind of experiment that I think would revolutionize physics. What it deals with has been thought of as the matter of interpretation: Are observers special? Are measuring devices classical, or are they also quantum systems? The main point of it is to show that the picture of a reality based on q-waves—which a lot of people call the many-worlds interpretation, but which I like to call the "everything is a q-wave" interpretation—is perfectly consistent with seeing definitive measurement outcomes. The beauty of this way of looking at things in quantum physics is that we don't need special observers to make sense of observations. Indeed, it's best to think of every physical system in this universe as being able to observe or be observed by any *other* physical system.

An illustration of an experiment in which one observer, Bob, couples to a quantum superposition, while another one, Alice, remains disentangled. By interacting suitably with the room in which Bob is, Alice can show that Bob is at some stage entangled with objects in that room (cat, poison, atom). Meanwhile, in each branch of that entangled state, the corresponding Bob sees a definitive outcome. This experiment is designed to show that there is no measurement problem in quantum mechanics.

What I intend to present is actually a simple variant of the Schrödinger's cat experiment. The idea that quantum mechanics applies to everything in the universe, even to us humans, can certainly lead to some interesting conclusions. It is this experiment that would conclusively demonstrate that there is no such thing as a measurement problem in quantum mechanics. It is also this experiment that tells us that any object is capable of playing the role of an observer—consciousness or complexity has nothing to do with it. Here we will see that such things are no longer just a matter of interpretation since we can actually test them.

REVISITING EINSTEIN'S COMPLAINTS

Quantum physics is well-captured by two of Einstein's complaints. His first was the famous statement "I don't believe that God plays dice with the universe." At the most fundamental level, all quantum events are unpredictable. Imagine sending a photon through a mirror that we call a beam splitter. You can think of it as just your normal sunglasses: there is a probability that the photon gets transmitted through, and there is a probability that the photon gets reflected. The result is that it "splits" photons into superpositions.

Einstein was dissatisfied that there was no deterministic way of predicting which of these two will happen during each run of the experiment. You get up in the morning, you have exactly the same breakfast, you make sure that all the conditions inside your lab are identical—the temperature,

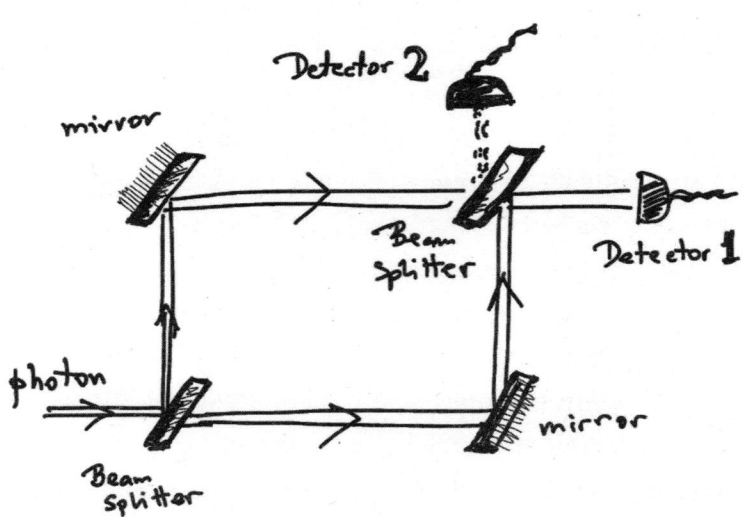

This is one of many ways in which quantum superpositions can be confirmed. The first beam splitter puts the photon into a superposition of two paths. The second beam splitter then combines the two paths into one output.

the humidity, anything else you can think of. You then press the button on your laser, and out comes a single photon. You run this experiment many times, always doing the same preparation, yet there's nothing in quantum physics that tells you which of the two detectors will click during each run of the experiment. Sometimes one detector will detect a transmitted photon, and other times the other detector will detect a reflected photon. All you can predict is that when you repeat the experiment many times, in half of the runs one detector clicks, while in the other half of the runs the other detector clicks.

Einstein's second complaint about spooky action at a distance (which I hope you are quite familiar with by now) can be answered with a beam splitter, too. We start with the same beam splitter, but instead of detecting the photon, we insert another beam splitter. The configuration with two beam splitters is known as an interferometer.

Now imagine that you run a photon through both beam splitters in sequence. If you check only the second beam splitter and not the first, which of the second beam splitter's two detectors should click at the end?

Einstein would have argued that the results should still be random. I mean, two random things one after another is still random, right? You can think of the first beam splitter as classically tossing a coin to decide what to do with the photon. If it's heads, it goes one way; if it's tails, it goes the other way. Then comes the second beam splitter, which is also like tossing a coin. You'd still expect to see heads half the time and tails half the time. In our experiment, Einstein would expect half of the photons on average to come out of the second beam splitter reflected and the other half to come out as transmitted.

Actually, quantum mechanics completely contradicts this reasoning! This kind of output never happens in the lab. Instead, every time you fire a photon, it always comes out of the second beam splitter in one of the detectors; the other detector never gets triggered. And this behavior can be fully predicted by using the formalism of quantum mechanics; the outcome is deterministic.

How can that be? This is a little bit like if I asked you to bet on me tossing a coin. You would surely reply, "I'm not going to bet on that; it's a random event, so there is a 50 percent chance that I'll lose money." Then I pull out another coin and I say, "Look, now I'm going to be tossing two coins." You would surely still decline to bet—even more so, as there is twice as much randomness now. But if tossing coins was like our quantum mechanical experiment, then you should say, "Yeah, I'm definitely going to bet now, since I know exactly the outcome of the two tosses!" This is just what quantum mechanics says: you can put two random things together and get a 100 percent deterministic outcome.

The only way we can understand this is by acknowledging that the photon really goes both ways through the first beam splitter. In other words, it exists in a superposition of both being reflected and transmitted. If we made a measurement at the first beam splitter, we wouldn't know which detector would click. But we *do* know that the photon is in a superposition: in one branch, the photon went one way, while in the other branch, the photon went the other way. The creation of the superposition itself is not subject to chance. These two definite paths go on to interfere with each other and produce a fixed, predictable outcome at the second detector!

This brings us to the point we've emphasized throughout the book—it's better to think of a photon as a wave than as a particle. When the two waves come to the second beam splitter, they cancel each other out at one detector, just like water waves do when one has a maximum amplitude and the other one has a minimum amplitude. On the other hand, they constructively amplify each other at the other detector, as water waves do when both have a maximum amplitude.

It's all about quantum things being q-waves. The bottom line is that somewhere in the game, you have to acknowledge that the photon really exists in both of these arms of the interferometer. And that brings us back to spooky action.

A STELLAR INTERFEROMETER

Imagine we have a telescope that is collecting photons coming from a distant star behind the Sun. Quantum mechanics says that every photon coming from that star is in a superposition of taking both routes around the Sun to enter my telescope. This situation is essentially a huge interferometer—there's a photon coming from a distant object, simultaneously taking both the right-hand path and the left-hand path around the Sun. It's a huge superposition across millions of kilometers.

If you decide to check which way the photon goes and you stick a detector in one path, you might not get a click there. That means there is no photon in that arm, which means that you "created" a photon in the other arm—the interferometer millions of kilometers away. And quantum mechanics seems to say that you've done that instantaneously. If you don't get a click in one place, then there is a photon that immediately appears at the other side of the universe. It sounds as though you should be able to communicate faster than the speed of light using this logic. This is precisely why Einstein didn't like it—it would seem to contradict relativity.

But this kind of communication doesn't work, and we know it because quantum physics is not (Einstein) non-local and there is no spooky action—everything travels at most at the speed of light, as postulated in relativity. Einstein needn't have worried at all.

Why, though, does this kind of protocol not work? Imagine that two people, Alice and Bob, want to communicate using quantum superpositions. Alice is sitting close to one of the arms of the interferometer and Bob close to the other one. Alice asks Bob, "Would you like to go on a date with me tonight?" And she wants a clear answer to this question, yes or no.

Bob inserts a detector into his arm of the interferometer. Say he gets a click, meaning that he detected a photon. That automatically means Alice won't get a click in her detector, which is positioned in her arm of the interferometer. Say that Alice getting no click means that Bob said no to her. And if Bob doesn't get a click, that means Alice will get a click, which is a yes to her question. It seems we have instantaneous communication between Alice and Bob, at least regarding this one bit of information.

What saves the day is the fact that God plays dice—Bob cannot know ahead of time whether he will get a click or not in his detector. The detection is an unpredictable, random event. So even though Bob would like to say yes to going out on a date with Alice, he cannot control which of the two messages gets sent. All Alice gets on average is a random answer, sometimes a yes, sometimes a no. That won't do, as Alice needs a clear yes or no.

A striking way of putting this is that God plays dice with the universe in order to save relativity from spooky action at a distance! As we've been saying, the randomness comes from the q-numbers. Therefore, we *need* the q-numbers in order to comply with relativity. As we saw, we could use an interferometer between Alice and Bob to communicate, but this is because the outcome in interference can be made to be deterministic. If the photon comes out from one port of Bob's beam splitter, he decodes this as a yes; otherwise it is a no.

The weirdness of quantum mechanics occurs beyond the level of photons. You can make any material object exist in many different states at the same time, and it would always be just a more complicated variant of the experiment we've been exploring here. How far have we taken this? We've confirmed superpositions with nuclei, atoms, electrons, and smaller molecules.

Einstein and his arch-opponent Niels Bohr argued at length over the meaning and nature of quantum mechanics—the one arguing that reality must ultimately be classical, and the other that there is no reality until we measure it.* Einstein's and Bohr's arguments were destroyed by Schrödinger, who showed that both positions are untenable. Schrödinger said that while we think of classical physics as applying to objects that are

---

* Some of these debates took place in Paul Ehrenfest's house in Leiden. Years later I had the privilege of being one of the few guests to stay at that house. It's owned by a private individual, and I think the Dutch government was desperate to buy it from him. He didn't intend to sell—I certainly would never sell a house like that! On the wall of one of the rooms on the top floor, which is where all these debates took place, you can find a framed picture with the signatures of all of the participants in these debates, a who's who of the foundations of quantum mechanics. It's an amazing piece of history, and I hope it stays preserved no matter who ends up owning the house.

large compared to atoms and slow compared to the speed of light, and while we think of quantum physics as applying to microscopic objects, the two domains cannot really be separated. After all, we're just an agglomeration of a bunch of small objects like atoms and molecules that are coupled via the electromagnetic field. Therefore, the quantumness of small objects can (and must) affect larger objects as well.

So where do you draw the boundary, Schrödinger asks, and why should quantum mechanics stop at the level of small objects? You can't be a naïve realist once you realize that the whole universe is entangled through ongoing interactions. And you also cannot hide behind the pragmatist view, because once a large object is confirmed to be in a quantum state, what does that mean? You can no longer say, "It's a weird small world, and this weirdness doesn't really apply to large objects." If it does, and everything really is quantum, then you simply have to rethink what this ultimately means. And, as I said, I think it means that the universe is made up of q-waves.

Hugh Everett, famous for his many-worlds interpretation of quantum mechanics that says that every possible outcome must exist, usually gets the credit for promoting the view that everything is quantum and that measurements are just entanglements between different quantum systems. Everett emphasized the relative nature of quantum observations, meaning that relative to my state of being happy the state of the cat is alive, while simultaneously there is another branch of the quantum state in which the cat is dead and I am sad. Many others would develop the idea further. So within each branch a classical reality is realized, while across the superpositions all outcomes may well be present.

Consider this variant of the iconic Schrödinger's cat thought experiment that Eugene P. Wigner came up with in 1961 and that David Deutsch of the University of Oxford elaborated on in 1986. It is Deutsch's extension that is key, because it contains an element that both Schrödinger and Wigner missed. I will try to use language that is as free of interpretation as possible and let you draw your own conclusions. I've personally witnessed all sorts of reactions from physicists upon hearing this: I've seen a conversion from the Copenhagen interpretation to the many-worlds interpretation, as well as "So what's the big deal with this stuff?" In my experience, the "So what?" reply typically comes either from those who truly understand quantum physics or from those who completely missed the point (the latter being more numerous than the former).

Suppose that a very able experimental physicist, Alice, puts her friend Bob inside a room with a cat, a radioactive atom, and cat poison that gets released when the atom decays. You might think I've included Bob because I still subconsciously cling to the idea that human observers are special and necessary to the whole undertaking. That's not it. The point of having a human there is that we can communicate with him. Getting conclusive answers from cats is not that easy.

As far as Alice is concerned, the atom enters into a state of being both decayed and not decayed, so the cat is both dead and alive but correlated with the state of the atom. Bob, however, can directly observe the cat and sees it as one or the other. This is something we know from everyday experience: we never see cats that are both dead and alive. In fact, despite many claims to the contrary, quantum mechanics says that we can never see a cat that is simultaneously dead and alive. That is exactly what happens when q-numbers entangle to other q-numbers—we get c-numbers.

To confirm this, Alice slips a piece of paper under the door asking Bob whether the cat is in a definite state. He answers, "Yes, I see a definite state of the cat." It is important that Alice does not ask Bob what that state is. The reason is that if she found out whether Bob saw the cat alive or dead, Alice herself would become entangled with her experiment and would not be able to perform the steps that are to follow the note. As long as she doesn't ask, all of Alice's q-numbers stay independent from Bob's (and the cat's and the atom's). Wigner actually missed this point. He imagined himself in the role of Bob and then imagined that he communicated his result to his friend. (It's for this reason that the experiment sometimes goes under the name of Wigner's Friend.)

It is logically possible for Bob not to see a definitive state and to say this. This fact in itself would be a refutation of quantum mechanics. However, this outcome seems unlikely, as it would also contradict classical physics. Nevertheless, this is worth bearing in mind, since all the experiments we will be looking into are tests of quantum physics, and quantum physics could ultimately fail.

So long as Alice doesn't make Wigner's mistake, the state of the whole system has changed from the initial state

|no-decay>[*] |poison in the bottle> |cat alive> |Bob sees live cat> |blank piece of paper >

to the state (as written from Alice's global perspective)

---

[*] The notation "|...>" is simply a way of representing quantum states (known as Dirac notation).

(|decay> |poison released> |cat dead> |Bob sees dead cat> +* |no-decay> |poison still in the bottle> |cat alive> |Bob sees live cat>) |paper says: "Yes, I see a definite state of the cat">

Those two definite states, live cat and dead cat, exist at the same time. I am assuming that because Alice's laboratory is isolated, every step leading up to this state is fully compliant with quantum mechanics. This includes the decay, the poison release, the killing of the cat, and Bob's observation—Alice has perfect quantum coherent control of the experiment. I will explain in the next subsection what this perfect control entails.

Note again that Alice does not ask whether the cat is dead or alive because that would "force" the outcome or, as some physicists might say, "collapse" the state. She is not like Bob in the Wigner's Friend scenario, who must communicate his observation; she is content observing that Bob sees the cat either alive or dead, and she does not ask which it is. Because Alice is content with not knowing the details of what Bob has seen, there is a remarkable implication: quantum theory holds that slipping the paper under the door was a reversible act. Reversible simply means that it could be undone. As I've explained, quantum physics conserves the overall information, meaning that all transformations can always be undone. She can undo all the steps she took, since each of them is just a fundamental quantum transformation (i.e., governed by the Schrödinger equation). This includes undoing Bob's observation, the cat's death, the releasing of the poison, and the decay of the atom. But the paper with Bob's answer remains. In other

---

* The "+" indicates that the two alternatives are in a quantum superposition.

words, the paper, just like Alice, does not get entangled with the rest of the laboratory and the experiment.

When Alice reverses the evolution, if the cat was dead, it would now be alive, the poison would be in the bottle, the particle would not have decayed, and Bob would have no memory of ever seeing a dead cat. If the cat was alive, it would also come back to that same initial state. Quantum interference is exactly a perfect reversal of what is described here.

And yet one trace remains: the piece of paper that says "Yes, I see a definite state of the cat." Alice can undo Bob's observation of the cat in a way that does not also undo the writing on the paper. The paper remains as proof that Bob had observed the cat as definitely alive or dead. As I mentioned earlier, I am saying this in an interpretation-neutral way. A many-worlds supporter would say that there are two copies of Bob, one that observes a dead cat and one that sees a live cat; a Copenhagen supporter or quantum Bayesian could say that relative to one state of Bob the cat is dead, while relative to the other it is alive. Either way, they ought to make the same predictions in this experiment (unless they think that quantum physics would collapse altogether).

THE CONTROL ALICE NEEDS TO HAVE

Let's spell out the details of Alice's engagement. The experiment is set up with an excited atom (which Alice can trigger from outside of the lab by pressing a button controlling a laser inside). The atom then spontaneously emits (a process fully governed by quantum mechanics). This superposition of emission and no emission then couples to the bottle with poison. Then the cat couples to poison (and no poison), and the assumption is that this process, too, is ultimately just a quantum evolution. All this is determined by the relevant q-numbers of the atom, the light field, the cat, Bob, and the interactions between them. At this stage, we have a superposition of two branches: in one Bob sees (seeing is also governed by quantum physics) a dead cat, and in the other he sees a live one.

But now comes the difficult bit. Given that the whole process is quantum, and that quantum physics is a reversible theory, Alice ought to be able to undo all of the processes so far. And this is the second half of the experiment. Here Alice needs to carefully engineer (in advance) all the steps that involve, first of all, the undoing of Bob's observation. All his excited neurons have to be de-excited, and the photons that entered his eyes telling him about the state of the cat have to be emitted backward and reabsorbed by the cat. Then the cat has to be returned to its previous state, before it interacted with the poison, after which the poison must be put back into the bottle. Again, all of these are in principle quantum actions governed by the same Schrödinger equation. Nevertheless, you get a sense of why this is very difficult in practice. The complexity lies in the fact that many things happen at the same time because all the objects involved are macroscopic (other than the atom and the photons), and all actions have to be done with great precision or else Alice won't achieve the interference she needs to confirm the existence of both branches.

Needless to say, this would be a stupendously difficult experiment, and it is questionable if we will ever have enough experimental prowess to execute it. We are talking about a nearly perfect, fault-tolerant way of conducting the reversal described above. Still, this is what quantum physics says would happen as far as its formalism is concerned.

And, as we will describe below, the logic of this experiment can be simulated with simpler systems. In the final analysis, all it takes is three qubits to capture what's going on (one qubit for the atom, poison, and cat together, one for Bob, and one for the piece of paper containing Alice's question and Bob's answer).

I present the circuit in the figure below. The simulation is easy to do with current quantum computers, as they can manipulate thousands of qubits with high fidelity.

Here is a startling conclusion for someone who believes that measurements have definite irreversible outcomes. Alice was able to reverse the observation because, as far as she was concerned, she avoided collapsing the state; to her, Bob was in just as indeterminate a state as the cat was. But her friend inside the room, Bob, thought the state did collapse. That person—or each person in each branch—did see a definite outcome; the paper is proof of it.

In this way, the experiment demonstrates two seemingly contradictory principles. Alice thinks that quantum mechanics applies to macroscopic objects: not just cats but also people

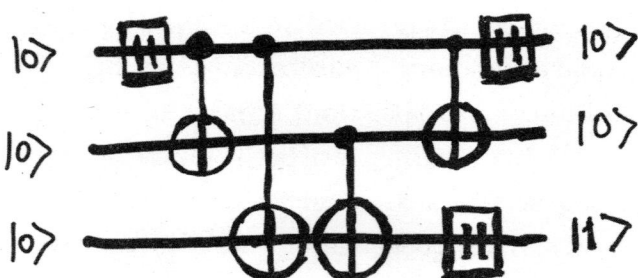

A simple network with three qubits showing the logical structure of the experiment with Alice, Bob, a cat, and a piece of paper. It requires only a handful of quantum gates (governed by the Schrödinger equation) to implement it.

can be in the quantum limbo. Bob, on the other hand, halfway through the experiment and before Alice's reversal, thinks that the cat is only either dead or alive—just as one would in the classical world we are used to.

I deliberately avoided interpretational jargon because this experiment cannot distinguish between different interpretations (nothing can—that's why they are called interpretations).* The experiment does, however, tell us whether quantum physics is valid at macroscopic scales or if there is a genuine collapse due to observation. If Bob collapsed the quantum state inside by seeing one outcome or the other (definitely dead or definitely alive), then the reversal would not be possible. Therefore, repeating this experiment a few times would tell us if the observation leads to a definitive collapse or not. For instance, sometimes Alice would get the outcome that the cat is dead (and Bob is sad), even though she would have taken all the correct steps to reverse things. This would presumably signal that Bob's observation of the cat was indeed definitive and irreversible, and that only one component of the quantum state had survived his intervention.

This experiment clearly shows that it's not just in relativity that different observers can see different things. In quantum physics, too, different observers see different physical situations. Alice sees an entangled state of Bob with the cat, while Bob sees one or the other outcome. Continuing this logic, quantum physics allows us to have yet another observer

---

* The diversity of quantum interpretations reminds me of Artur Ekert, one of the inventors of quantum cryptography, who said that the best way to agitate a group of physicists is to buy them a bottle of wine and mention interpretations of quantum mechanics. He said that he'd done this on many occasions only to discover that the number of viewpoints often exceeds the number of participants.

who is experimenting with Alice while she is experimenting with Bob. A quantum universe is a ginormous (and potentially infinite) entangled network of such nested observers. And all are consistent, even though their factual states differ.

Doing such an experiment with an entire human being would be daunting, but physicists can accomplish much the same with simpler systems. This is important, as it shows that any qubit can be thought of as an observer.

We can take a photon and bounce it off a mirror. If the photon is reflected, the mirror recoils, but if the photon is transmitted, the mirror stays still. The photon plays the role of the decaying atom: it can exist simultaneously in more than one state. The mirror, made up of many atoms, acts as the cat and as Bob. Whether it recoils or not is analogous to whether the cat lives or dies and is seen to live or die by Bob. The process can be reversed by reflecting the photon back at the mirror. If the photon always comes out the way it came in, we confirmed that it was in a superposition after the mirror and before the reversal. Otherwise, there was a collapse somewhere along the way (needless to say, there is no collapse if things are done properly in actual photonic experiments of this kind), and that collapse would lead to the photon coming out on the other side of the mirror.

We can do similar experiments with atoms and molecules, during which we entangle them and subsequently disentangle them. Again, no collapse has ever been recorded.

In developing this devious thought experiment with Alice, Bob, and the cat, Wigner and Deutsch followed in the footsteps of Erwin Schrödinger, Albert Einstein, and other theorists who argued that physicists had yet to grasp quantum mechanics in a deep way. For decades, most physicists scarcely cared because the foundational issues had no effect on practical applications of the theory. But now that we can actually perform many of these experiments, the task of exploring the full extent of quantum mechanics has become all the more urgent.

To me, the validity of quantum physics at the macroscopic level naturally suggests the "everything is a q-wave" picture. The simple fact demonstrated here—namely, that from one perspective (Bob's) we can have definite outcomes while at the same time everything remains in a quantum coherent state from a higher perspective (Alice's)—is a definitive bit of evidence telling us that the whole world can be described consistently using quantum mechanics.

GEOMETRICAL OPTICS ANALOGY

I'd like to explain why "many-worlds" is not the most appropriate name for the view I am arguing for here. For this, I will use an analogy with geometrical and wave optics (both of which are classical descriptions of light). The parallel between classical optics and quantum physics is apt, and it is no accident that it led Schrödinger to his famous equation.

Historically, light was thought to travel in straight lines in a uniform medium. It curves only when it goes from one medium (e.g., air) to another (e.g., water): it reflects when going back to air and refracts when it enters water. Such laws of behavior constitute geometrical optics and explain how mirrors and lenses work, how telescopes and microscopes are to be designed, and how rainbows are formed.

But geometrical optics cannot explain why light, even when traveling in a uniform medium, disperses. All beams of laser light eventually start to broaden. Also, like sound, light bends around corners. To explain this, we need to treat light as a wave phenomenon. Doing so helps us account for all the wonderful phenomena of diffraction and interference that we observe everywhere around us.

Mathematically speaking, the wave description contains geometrical optics as a special case. When the wavelength of light becomes small, the wave optics become approximately geometrical. In the limit that wavelength approaches zero, all interference and diffraction phenomena vanish. You can no longer explain, say, Newton's rings (concentric circles that appear when light bounces between a convex glass surface and a flat glass surface).

The interesting observation is this: if you hold to geometrical optics but allow those straight lines to meet and interfere like waves do, you are able to explain quite a lot of (but not all) diffraction phenomena. The Fraunhofer limit defines the conditions under which this approximation scheme breaks down and the conditions under which this approximation works.

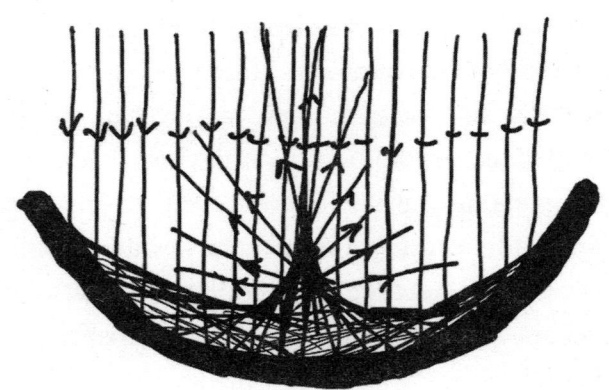

Geometrical optics is capable of explaining the beautiful reflection patterns on the surface of coffee we get when light falls sideways on a cup. This does not require us to think of light as a wave.

In Schrödinger's wave mechanics, we have something analogous. Every particle is a wave, but as the wavelength of the particle becomes small (one could think of its mass becoming large), we recover the classical behavior of Newton. This was the key insight that led Schrödinger to his own formulation of quantum physics.

The point I want to make now is that the many-worlds view of quantum physics could be thought of as the "Fraunhofer limit" of the "everything is a q-wave" picture. The many-worlds interpretation says that we have a superposition of classical worlds, a structure that is exactly like the lowest level of interference of straight paths of geometrical optics. Each straight line is akin to a classical world, and in the many-worlds view they interfere with each other. However, the full picture—when we treat the whole universe quantum mechanically—is much broader than that.

In wave optics, this more complete picture is called the Fresnel regime. Here we have no geometrical straight lines, and everything has to be treated like a wave. The same is true in quantum mechanics, where, in the most general situations, we cannot talk about a superposition of classical things. We have a q-wave, but this wave cannot be decomposed into fully distinguishable classical worlds. There is nothing classical here at all, only the simultaneous presence of all q-waves. This is why I think that the name "many-worlds" is inappropriate—it's still a remnant of the classical jargon and it addresses only a special case of the total reality, and until we get rid of it, until we fully embrace quantum mechanics, it's going to be hard to make genuine progress. The portals into the new reality are all, each in its own way, a consequence of this fact. In the final chapter, I will explore the wider consequences of this fact and ask if our perception could be augmented to become attuned to it.

ONE MORE TWIST

If this wasn't enough to excite you, let me tell you about one more twist in the tale of Alice and Bob. Halfway through Alice's experiment, she knows that if quantum physics is upheld at that level, then Bob is entangled with the cat, and that there are two copies of Bob, one seeing a dead cat and the other a live cat. But does Bob know that there are two of him? Well, not at all—in each branch of the superposition, Bob feels as though there is only one world. That is, unless Alice tells him otherwise!

Before reversing the evolution (and after Alice receives a communication from Bob that he sees a definitive outcome), Alice can actually communicate with Bob again. This time, however, she doesn't have a question. Instead, she says to him (by slipping another piece of paper under the door): "From your reply, I know you see a definite outcome, but I am now telling you that you are nevertheless in a superposed state of seeing both outcomes. More precisely, there is a version of you (or of your consciousness or whatever constitutes you) that sees the cat as dead and another one that sees the cat as alive."

Even better, if Bob is himself a quantum physicist, Alice could just write down the equation describing the state of the laboratory on the same piece of paper. This equation would be the same as the one we wrote on page 113.

If Bob trusts Alice (she is, after all, both a great friend and an amazing physicist), he might be shocked. He might say to himself, "I see a definitive outcome, so how can I still be in a superposition?"

The answer to this conundrum is, of course, that Bob is not in a superposition. Rather, he is entangled with the cat and the poison and the decayed atom, exactly as above. And being maximally entangled with something means not being in a superposition yourself.

Here we also see that in each branch Bob is unaware of his observation in the other branch, but the observation that he is not aware of can indeed affect his future (in each branch) if Alice decides to reverse everything and unify the two branches. This is why I keep insisting on the fact that the unobserved outcomes could still affect our future in the quantum universe.

So after Alice's second communication, Bob knows he exists in two different "worlds" (or rather, each version knows about the existence of the other), yet each of the two versions of him feels as though it is safely operating within one world only. Note also that even though I am promoting the picture in which everything is quantum mechanical, all other interpretations of quantum physics will have to agree on the outcome (unless quantum physics fails, in which case we need a new theory and not another interpretation). For instance, a Copenhagen person would insist that there is always a need for an external classical context. They would admit that everything in the lab, the atom, the cat, and Bob are in a superposition, but that Alice herself provides the needed classical context. I am, on the other hand, arguing that Alice could be quantum, too, and that we never need that ultimate classical domain.

In fact, Alice can perform measurements to confirm that Bob is in the entangled state without collapsing the state and then send the experimental results to Bob to dispel any lingering doubts he might still have. Of course, Bob would have to trust that she performed the relevant experiments and that the results he has received from her are genuine.

And trust is exactly the right idea here: the logic behind being able to confirm entanglement while not destroying it is the same as in the famous conundrum involving two brothers, one of whom always tells the truth and one of whom always lies. Each of the brothers is guarding a door, and you have to decide which of the doors to pass through. The major wrinkle is that one leads to a certain death, whereas the other one leads to a life of prosperity and happiness. The brothers know which door is which, but you are allowed to ask them only one yes-no question to try to find out. After that, you need to walk through one of the doors.

If you've never encountered this kind of puzzle, your first thought would probably be that this task is impossible. How can a single question tell us which door leads to prosperity and happiness when we don't even know which brother is a liar and which not? Well, it is possible to do so, and the amazing thing is that at the end you will know which door is the right one, even though you still won't know which brother is which!

This can only mean that you need to ask one brother about the other one. And it doesn't matter which one you ask, since the situation is completely symmetric. The question to put to either of the brothers is this: "What would your brother say if I asked him whether this door leads to death?" Suppose the answer is "My brother would say that yes, this door leads to death." If you asked the lying brother, that means that the door you're indicating does *not* lead to death. If you asked the truthful brother, then it would mean that his lying brother would have said "Yes, that door leads to death," which in turn means that no, that particular door doesn't actually lead to death.

124

Therefore, if the answer to your question is "My brother would say yes," you should take the door. If the answer to your question is "My brother would say no," it means you should take the other door.

The critical thing about the puzzle is that the behavior of the brothers is correlated. The notion of being correlated simply means that if we know the status of one of the brothers (a liar), we would immediately know the status of the other one (a truth-teller). The truth about the doors is revealed precisely because of their correlation (sometimes called a negative correlation because they are the opposite of each other), even though we don't know which is which.

The same goes for entanglement in the above experiment. After Alice tells Bob he exists in two worlds, she offers him a bet. She tells him that she will now perform a reversal of the experiment and thereby interfere the two possibilities. She says to him: "What will you see at the end when I am done with the reversal?" We seem to be facing two logical possibilities. One is that Bob will be in the same state as at the beginning of the whole experiment (and the cat will be alive, the poison in the bottle, and the atom not decayed). This is what I argued quantum physics would lead to. The other is that he will see the cat dead, the poison out, and the atom decayed. But how can this second possibility happen?

It cannot, unless there is a collapse of quantum physics due to Bob's observation. If there were a collapse, after Alice's reversal Bob could, in principle, end up in the dead-cat branch. *That* would be a proof that quantum physics cannot describe the overall experiment.

In the analogy with the two brothers, failing to reverse things to the initial state would, in Alice's experiment, mean choosing the wrong door to go through. In fact, this analogy presents an excellent metaphor for the main message of my book. We have been choosing the wrong door as far as our physics questions are concerned precisely because we have not thought of quantum physics as applying to everything.

Alice's experiment is a test of the consistency of the all-encompassing quantum picture. If Bob believes Alice (as far as the state of affairs is concerned) and quantum physics is upheld, then he should bet on seeing the cat alive at the very end (even when he is in the branch where he observed the dead cat previously).

ON THE VERGE OF IRRESPONSIBLE

I'd love to see a full psychological analysis of the concept of personal identity in quantum experiments of this kind. Bob splits into two Bobs, each of whom remembers all the things that Bob did before the split, but after the split each has new experiences different from the other's. On top of it, the two Bobs are then united by Alice into a single Bob, which therefore erases both of their memories and returns them to the state of Bob before the split. Even the visionary philosopher Derek Parfit, who considered the question of personal identity in various science fiction scenarios, did not contemplate this crazy quantum possibility.

But could this kind of experiment ever be performed with a human being? I have no idea, but imagine something like the following (I am now speculating, and on the verge of being irresponsible). There is a famous two-dimensional line drawing of a cube first brought to light by Swiss psychologist Louis Necker. Even though there are many logical interpretations of this picture, the brain does not see any ambiguity in this two-dimensional drawing. In fact, we only ever see two of the possible images (with one of the surfaces at the front or at the back). Most people see one of these first and then, after a few seconds or so, see the other one, thereafter continuing to flip back and forth between the two.

Suppose now that the two images of the Necker cube are stored in our brain as two distinct quantum states. Admittedly, these states could be very complex, in the sense of involving many atoms and interactions between them. The mind then switches between the two physical states, which is like a logical operation of a flip from 0 to 1 and back from 1 to 0. The question is: could it be that this switching process is actually quantum mechanical and that during the transition our brain is actually storing a superposition of the two images?

If so, maybe this could give us a small window of opportunity to be able to do something like Alice did. Perhaps we could confirm that Bob sees a definitive version of the Necker cube and then undo this observation (providing we understand that his perception is stored), thereby demonstrating that Bob has literally been, quantumly speaking, in two minds at the same time. This would not mean that quantum physics was necessary for perception, and even less so for consciousness. However, it would constitute the most remarkable example of a macroscopic quantum effect to date.

Of course, even the original experiment is challenging in practice. The reason is that it relies on the sort of communication between different observers in the experiment that only humans can do. Other animals do communicate, but—at least at present—we don't understand these languages reliably enough. You might replace Bob with, say, any computer.

In fact, my impression is that it's more likely that a very simple artificial intelligence (AI) system will actually be the first to undergo this kind of experiment, in which we would communicate with *it* and ask *it* questions about its state. After that we might try to put simple life, such as viruses or bacteria, in a superposition. (You could imagine that is something we might like to do with the virus that causes COVID-19, if only as revenge, but such a thing is probably two orders of magnitude too large given present technology.) In principle, you can do these experiments with two microbes confined between two mirrors, such that a split photon would excite one and not the other. Of course, you cannot communicate with bacteria and ask them whether they feel they're in a definitive state of being excited or not, but you can confirm from the light they emit whether they are entangled or not. This is typically done by shining laser light onto the bacteria and seeing whether the emitted light exhibits special correlations that would be indicative of entanglement.

FURTHER DOWN THE RABBIT HOLE

What if there is another observer—call him the Mad Hatter— who is outside Alice's laboratory, and so he can control Alice, who in turn can control what happens to Bob, who in turn is observing the cat? The Mad Hatter is now the person who must ensure that quantum interference takes place with Alice, Bob, and the rest. This requires him to be able to, in the second part of the experiment, undo whatever happened in the first part (as we saw when Alice had the control).

You can follow the same logic as before, and you'd come to the same conclusion as we did before. If Alice figures out what's happening with Bob, she can communicate this to the Mad Hatter and say, "Look, Bob has made an observation and the cat is alive," but as soon as she does so, both she and the Mad Hatter are part of Bob's world, in which there is no question of the cat's state. However, if quantum physics applies, then both Alice and the Mad Hatter have now joined the superposition, and each exists in both of the branches. From the point of view of Lewis Carroll, who has not interacted with them in such a way as to get entangled, there must be a world in superposition, where there is also the bad news that the cat is dead.

Such a pattern of regress has astonishing implications for the nature of the reality we experience. Consider a painter who wants to faithfully paint the forest he is standing in, down to the most minute detail. When he finishes, he realizes that one important bit is missing: himself. So he decides to add himself, but then of course realizes that the painter painting the first painter is missing—and so on ad infinitum.

This is very similar to how, in the picture we've been painting, a definite reality emerges through a sequence of entanglements that you get from interactions that are completely quantum mechanical: between the photon and the poison, the poison and the cat, the cat and Bob, Bob and Alice, Alice and the Mad Hatter, and so on. As with the somewhat paranoid painter, it's not a completely faithful image of reality, since there is always another observer missing. But at the same time, this does not prevent the scene presented from being definite. In quantum physics, no observation is ever final, and they are definitive only within realities that exist in overall superpositions.

Whenever I talk about these nested realities, I inevitably get asked what it would feel like to be in another universe, where (for example) the cat was dead. I suppose the answer must be that I would be sad! But the question leads to another fascinating fact about the nature of time. It turns out that, quantum mechanically, to ask what it is like to live in a different universe is the same as asking what it would feel like to exist at another time. Two physicists, Don Page and William Wootters, showed this in the 1980s in a paper titled "Evolution Without Evolution." They suggested that the universe at different times is really just different quantum universes. Nothing really evolves, and everything that will happen has already happened: it's all sitting simultaneously in a "block" universe that contains all possible things that can happen. In this case, the components of the entangled state are just different times at which the universe exists. This, believe it or not, is a fully consistent way of thinking about quantum mechanics. Seeing time as a form of entanglement leads us to the same dynamics as described by the Schrödinger equation: one part of the universe changes in relation to the rest of the universe, while both of them taken together do not change at all!

Chapter 5

QUANTUM EXPERIMENTS WITH TIME

We now approach the second portal through which we can test new physics. This path may look rather different from the previous one, but this is only an illusion. Despite their unique twists and turns, all of the paths I explore in this book lead to the same destination.

The ideas explored in this chapter are intimately connected with the observation that different times are just like different realities in space. And the key principle behind this is the coherent control of quantum information. Any two distinguishable states of affairs can always be put into a quantum superposition.

ONE SECOND PER SECOND

There is an old book, *An Experiment with Time*, written in 1928 by English engineer John Dunne. It is an inspired account of the writer's struggles in trying to scientifically explain the phenomenon of premonition, for which he offers many personal accounts as well as those of other people.

Now, one doesn't necessarily have to believe in premonitions in order to enjoy Dunne's theoretical account of it. He ends up arguing for an infinite number of temporal dimensions. The first dimension of time is the one we normally think of—the one that merges with space into the familiar spacetime of relativity.

However, if we want to talk about the flow of time—which Dunne maintains is necessary to explain premonition—then you need another dimension of time with respect to which you measure the speed of the first one. In other words, the speed of time in the first dimension of time is one second per second as measured in the second dimension of time.

You can now probably see where Dunne is going. First of all, we cannot stop at the second time because its own flow requires another, third, dimension with respect to which we would measure its rate of passing. And so on. Dunne fully acknowledges that this leads to an infinite regression, but—unlike most philosophers—he does not consider this to be an obstacle or a shortcoming of his theory.

Dunne's idea is that premonitions are our mind's leaps into the higher dimensions of time, within which we can move faster than in the first temporal dimension. In other words, the unit of time in the second dimension could be shorter than the unit of time in the first dimension. This is what allows us to "see" the future faster than it happens in the first dimension of time.

It's obvious that Dunne was influenced by both relativity and quantum physics. However, there is another notion in physics, that of a wormhole, which he could not have been influenced by because it came much later. The idea is that there are tunnels in spacetime that allow you to go between two different points faster than if you had traveled through "normal" spacetime. Had this notion existed at Dunne's time of writing, perhaps he wouldn't have needed infinitely many dimensions of time.

There is much to be skeptical about when it comes to Dunne's logic, entertaining though it is. However, we have already seen that, in relativity, objects traveling at different speeds measure different amounts of time. In this sense, relativity already contains infinitely many possible rates at which time could flow!

On top of that, we have also seen that in quantum mechanics, different observers could see different factual states of affairs. In one branch, Bob sees a dead cat, while in another, a different copy of him sees a live cat. And from the outside, Alice still thinks of the two as being in a superposition, so both outcomes are true at the same time. In that sense, quantum mechanics also allows for the existence of multiple times, since each copy of Bob could have a clock that showed a unique time in each branch, and both these times could be different from that of Alice's clock.

The big question is: what happens to time when we take both relativistic and quantum effects into account? Here we will be measuring time using atomic clocks. For all practical purposes, one tick of an atomic clock is measured by the time it takes an electron within the atom to make a transition between its lowest energy level and a specifically chosen higher energy level. Typically, it takes an electron a tenth of a millionth of a billionth of a second to do so, which sounds like a pretty short time interval. It is indeed short according to our everyday standards, but it is worth remembering that the time differences we want to measure, such as the difference between the rate of time at your feet and the rate of time at your hip, are also exceedingly small.

It is also worth mentioning that any process extended in time is sufficient to serve as a clock. For instance, the duration of the adenosine triphosphate (ATP) cycle in the cell mitochondria could serve as a clock in principle. However, such a clock is difficult to initialize, read out, or reset. Also, the natural smallest temporal steps in the ATP cycle are usually much larger than those of atomic clocks (we are talking milliseconds). So we use atomic clocks simply because they give us the best temporal resolution available. Relativity tells us that the amount by which an atomic clock slows down (or speeds up) must be the same as the amount by which any other temporally extending process is affected, too.

WHOSE TIME IS IT, ANYWAY?

The first interesting set of experiments that we could perform with today's technology but have not yet done so has to do with what happens when the principle of quantum superposition meets the relativity of time. These tests provide a bridge to ultimately testing the full quantum nature of gravity, our portal number three.

First, let's start with the special relativistic effect described by the twin paradox: the fact that time slows down when we move with respect to those that are stationary. The twin paradox is exaggerated for effect—one experiment we cannot perform with today's technology is sending a twin, or even a ping-pong ball, at four-fifths the speed of light—but although the effect is very small at the speeds of everyday life, it still exists. So if you were to fly from London to Brisbane and back, you would be about a ten-thousandth of a second younger than all the Brits who stayed behind. In 1971, physicists Joseph Hafele and Richard Keating took atomic clocks—capable of losing no more than one second every thirty million years— on a commercial jet, flying first west and then east around the globe before returning to their laboratory in Washington, DC. There they compared the time on their timepieces to a set of clocks that had remained static and found that the disagreement was exactly as Einstein's twins would have experienced.

The first quantum experiment with time would be to take an atomic clock and put it into a superposition of two different locations. Then in one of the locations we take the ticking atom on a journey away and back, while the other state remains static. When the two paths are brought back together, we have one clock, located in one and the same place, but it now is in a superposition of two different times! It is younger and older at the same time (here things get confusing as far as language, but "at the same time" means according to the time of the person performing this experiment—the clock itself is "quantum confused"). This superposition of times can be tested and confirmed experimentally like any other superposition. But the concept is intriguing since it adds a quantum twist on the relativity of time—and still only within the special theory of relativity (as opposed to the theory of *general* relativity).

The key idea here is that if something can move at two different speeds, then we can in principle combine these into one superposition. This is the coherent quantum control principle in action. It goes without saying that this experiment is non-trivial and the degree of control we might need is high—however, it is still nowhere near what we needed in the Schrödinger's cat variant.

Note another interesting phenomenon already alluded to above. Time is not just relative in relativity! It is also so in quantum physics. In the above experiment, one branch of the superposition shows one time, the other branch another time, and the experimenter performing the experiment has still a third time! If we could do this with a human, then their ATP

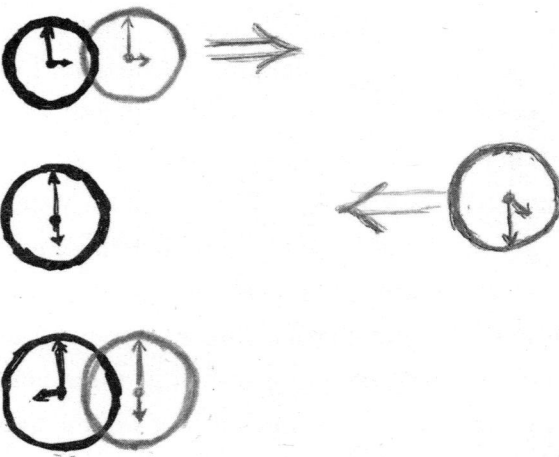

Making a superposition of a clock that stays in one place in one branch and travels at a certain speed in the other one. At the end of the experiment, when the two branches are reunited, we have one and the same clock, but in a superposition of two times.

cycle would take longer in one branch than in another. And this would ramify down through everything about their physiology, so they would be in a superposition of tired and rested at the same time. But let's not get ahead of ourselves . . .

If you are getting a sense of déjà vu now, there is a reason for that. The situation is the same as the "observing the observer" scenario with Bob in a superposition and Alice controlling the experiment from the outside. The conclusion we are now reaching is that this situation arises when measuring any observable, not just cats that are dead or alive. It potentially applies to space, time, momentum, energy, and so on. They are all relative, quantum mechanically speaking.

SUPERPOSITION OF GRAVITATIONAL DILATIONS

As we saw, general relativity, not just special relativity, warps time—time is slower for a person closer to the Earth relative to someone farther from the Earth. This, too, has been measured with our atomic clocks. You can watch YouTube videos in which an atomic clock is gradually raised by half a meter or so, causing its last two digits to change. Our atomic clocks are sensitive enough to measure the time difference between your hips and your feet, and this is exactly the spatial extent of the superposition required of an atom clock. It is a challenging experiment, but it is certainly doable. The second experiment would therefore be to put an atomic clock in a superposition of being at two different heights in Earth's gravity and again create a superposition of different times. Here, the superposition of different times would be caused by gravity, not by motion.

Consider a fascinating variant of this experiment. Put an atomic clock in a superposition of two gravitational fields for half the time of the total experiment. Then, for the second half of the experiment, reverse the two states so that the branch that was higher during the first half of the experiment will be lower and vice versa. By the experiment's end, we might expect the two clocks to show the same time. After all, each was delayed by the same amount with respect to the other for half the time of the total experiment. However, are things reversible when quantum physics and relativity are combined? This is exactly the feature that this kind of experiment would be testing. Most of my colleagues would probably bet that this experiment would be successful. We would confirm that there is a quantum superposition of different gravitational time dilations. But, as always in science, this box is important to check off because, in the unlikely event of the experiment failing, this would have huge consequences for how we approach the task of quantizing gravity.

It is even possible to superpose a particle undergoing two inertial motions: in one branch it is inertial according to Newton (constant speed), while in the other branch it is in free fall in Earth's gravity (accelerating) and therefore inertial according to Einstein. The two branches would interfere as shown in the following picture. The interference reflects the fact that we are superposing the same particle, but in two radically different states of motion. My friend, the Israeli physicist Ron Folman, has done some preliminary experiments to verify this prediction by Marletto and myself.

And now we shall enter a domain that seems to belong to science fiction, but which I assure you modern physics tells us is entirely possible.

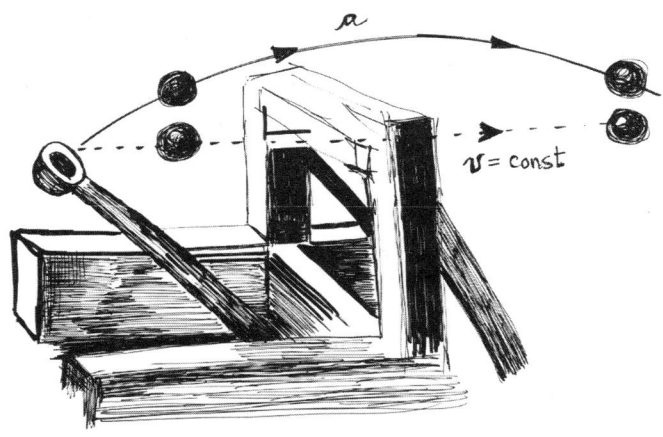

A quantum version of Galilean experiments in which he was launching projectiles in different ways to test how objects move in Earth's gravity. Here we have a quantum catapult (which is a sort of a beam splitter) that shoots out a superposition of a particle undergoing ballistic motion in Earth's gravity in one branch, while in the other branch the same particle is moving at a constant velocity. This is a superposition of acceleration and uniform speed, which in quantum physics interfere to reveal the difference between these different motions.

THE QUESTION OF IRREVERSIBILITY

Explaining the fact that virtually all processes in the macroscopic domain take place in one direction but never backward has eluded us ever since the days of Newton. A hot cup of coffee always cools down to reach the temperature of its surroundings. We (unfortunately) never observe the opposite process, by which a cold cup of coffee spontaneously heats up and becomes drinkable again. Without microwaves and other gadgets designed to provide energy, this kind of thing never happens all by itself.

This famous arrow of time presents a problem for physics, since—at the microscopic level—each atom that composes the cup of coffee and its surroundings can move equally easily in any direction, forward and backward and sideways. If we follow the motion of each of the atoms as the cup cools down and then reverse each of the atomic trajectories, they would all end up going backward and collectively undoing the cooling (thereby heating the coffee up). But this never happens spontaneously.

Why not? Some people would say that this impossibility is behind the famous second law of thermodynamics. One cannot get anything useful out of equilibrium, meaning a state in which everything is at the same temperature, without investing some extra work. For instance, we need to supply electricity to heat up the stove in order to reheat our cooled-down coffee. Yet nothing in the microscopic laws of physics tells us that this must be so.

The second law can be expressed in terms of entropy, a physical quantity that roughly corresponds to what we think of as disorder in everyday life: the entropy of a closed system (i.e., one that is insulated from its environment) never goes down. The entropy typically increases until it reaches its maximum, at which point the system is in a state of equilibrium. However, in the microscopic domain, we again have a problem accounting for the entropy increase. Our microscopic laws, those of quantum physics, are perfectly reversible, and this means that they should preserve entropy. The entropy of a closed system obeying laws of quantum (or classical) physics always stays the same—in apparent contradiction with the second law of thermodynamics.

Now, there are two ways one can argue for irreversibility from the underlying reversible laws (both arguments are due to Ludwig Boltzmann, the man who dedicated most of his career to this problem). Loosely speaking, one is based on statistics, the other on complexity. Even though Boltzmann only used classical mechanics, both can be applied to quantum mechanical systems, too.

According to Boltzmann, the process of cooling can be understood by analyzing collisions of individual atoms. When two atoms collide, their states become correlated. Boltzmann tells us that when each of these two atoms now interacts with two *new* atoms, the correlations established in the first collisions can be neglected. This neglect of the correlations is what leads to the entropy increase. Boltzmann called it the "assumption of molecular chaos" (actually, he used an incredibly long German word meaning the same thing, but this is not relevant here). In fact, the difficulty we have in doing the Schrödinger-like experiment or building large-scale quantum computers is precisely there because we have to battle against the existence of the underlying molecular chaos.

All fine and good, until the atoms that interacted initially come to interact again. French mathematician Henri Poincaré tells us that this must happen sooner or later if indeed the underlying dynamics (classical or quantum) is reversible. Here, the correlations that were initially established do become important and can even be undone in the next collision (upon which, if we buy Boltzmann's logic, the entropy would go down). The assumption of molecular chaos can therefore be justified for a while if the system under consideration has many atoms, since it might take a long time for a collision of the same two atoms to recur (thus we wouldn't count on it when it comes to quantum technology). But in the long run, atoms must reencounter each other, and so Boltzmann's assumption cannot solve our problem.

141

Boltzmann offered a second justification for the arrow of time when corresponding with his countryman Johann Josef Loschmidt. Loschmidt said that if all the velocities of all the molecules were reversed at some point during the dynamics, the system would then return to its initial state (the act of reversal is known as the Loschmidt echo). So it cannot be that the entropy always increases, since upon reversal it would have to go back to its initial value. Boltzmann's reply to this objection, sent on a postcard, was apparently "Go ahead and reverse them!" This is meant to imply that the complexity of tracking all the atomic motions and then performing the reversal of each atom's velocity is an exceedingly difficult (maybe even impossible) task. In other words, all the king's horses and all the king's men couldn't put Humpty Dumpty together again. Remember, it is the Loschmidt kind of reversal we need to be able to perform to undo the observation that Bob made of the cat. Well, it is possible that we would be in the same position as Humpty Dumpty.

So there we have it. The arrow of time in the macro world is due either to the erasure of microscopic correlations or to our (fundamental or otherwise) inability to perform a complete reversal of complicated macro dynamics. When I say "our inability" I don't mean to imply a special place for observers. But I am simply asking whether—when all the rules of the game are allowed—macroscopic processes could be undone. But is that all there is to it?

Many physicists are unhappy with either of the explanations of the second law. After all, the first law of thermodynamics, which is a statement of the overall conservation of energy in a closed system, is perfectly compatible with micro dynamics. There is no need to make extra statistical arguments

Erasure & Entropy

Two atoms colliding. If we lose the information about which atom is which after the collision, the overall entropy will increase. This is one of the explanations for the origin of the second law of thermodynamics according to Ludwig Boltzmann.

or invoke the complexity of performing certain operations in order to explain energy conservation. Energy is always conserved. Period. In quantum physics, as we've seen, energy is conserved in each branch of a quantum superposition.

Why can't the second law be just like that? One possibility is that the second law is simply not as fundamental as the first (I can hear Rudolf Clausius, Lord Kelvin, and Arthur Eddington spinning in their graves). It only happens to be true when we have large enough systems consisting of many constituents, and only then does irreversibility emerge from the underlying reversible laws (i.e., as we sprinkle them with a bit of statistics and add in some complexity). Many physicists would probably agree with this view of the second law (yours truly, guilty as charged).

However, my colleagues Chiara Marletto and David Deutsch think otherwise. Marletto, in particular, maintains that it is possible to phrase the second law such that it applies to

143

objects at all scales, not just at the macro scale. But given that it seems that total entropy cannot increase (or decrease) in quantum physics, how can this be?

Enter constructors. Constructors were introduced by John von Neumann to generalize computers to machines that are also capable of replicating. We can think of any physical process as a task that some constructor—any entity that can cause some particular transformation(s) and retain its ability to do so again and again—is able to execute. Von Neumann suggested using constructors to colonize Mars, but actually all living systems can also be thought of as constructors with different sets of repertoires (for instance, humans can perform a larger set of tasks than spiders can). Keeping to thermodynamics, a typical thermodynamic constructor would be a chemical catalyst in the sense that it enables a reaction to occur that would not be possible without the constructor. But, just like with some chemical reactions, which naturally run only in one direction, it could be that for a given task there is no constructor capable of performing its reverse (this is lucky for us and other living systems, since we certainly don't want our metabolism to suddenly start going backward).

In this way, Marletto and Deutsch would say that the second law could be put on an equal footing with the first one— namely, that both are universal and apply to everything. The constructor-based second law says that for some tasks there is a constructor in the forward direction, but it is impossible to have a constructor in the reverse direction. Thanks to the constructor twist, this formulation is, surprisingly, compatible with the reversible laws of quantum and classical dynamics. Remember, a constructor is not an observer or an agent; it's simply a physical, even inert, system that can perform transformations based on the laws of physics over and over again.

Our third test of quantum effects on time is possibly the weirdest, and some proof-of-principle experiments have already been done. If this test is successful, it could possibly leave us with the constructor-based formulation as the only viable one. As we've just seen, quantum physics tells us that any process happening in one direction can also happen in reverse. We can take an atom in its lowest energy state and excite it by shining some light on it. Once it is excited, we can then shine some more light on it and bring it back to the ground (lowest energy) state. Both processes can occur deterministically—that is, they can be done 100 percent efficiently in the laboratory.

Adding some quantum physics to this, we could superpose the two processes, one going in the forward direction and the other going in the backward direction. You may think, *Of course!* But this is actually quite weird: it is a phenomenon that goes under the name "superposition of causal orders." This is because which way time flows tells us which things are causes and which things are effects. First we press the remote control, and *then* the TV switches on. The former is the cause of the latter, and the latter is the effect of the former. Recall that in relativity, events can happen in different orders. This already sounds weird, but when you add quantum mechanics to it, it becomes even more so. Here we would like to make a superposition of two spacetimes such that the order of events between branches is reversed. And if this structure fails to be possible at some level, then there is a fundamental question as to why this is so. This is why this constitutes a possible path to new physics.

IMAGINARY NUMBERS

One way to think about relativity is that its whole point is to upgrade time to a fourth dimension (in addition to the three spatial ones) but with a twist. That twist is to use imaginary numbers to describe it. The reason, as you recall, is that in four-dimensional spacetime, time, as the fourth dimension, behaves differently than space does. The interesting thing about imaginary numbers, like the square root of -1, is that they were first described by the Renaissance polymath Gerolamo Cardano to solve equations such as $x^2 + 1 = 0$. He probably thought of the imaginary numbers as just a clever mathematical device and nothing more than that (the very name "imaginary" suggests that they are not as real as the real numbers). Everyone else thought the same, and the imaginary numbers continued to have this reputation as inferior for about four hundred years. But it is worth remembering Galileo, who said that the book of Nature is written in the language of mathematics. It took two major revolutions in physics to actually make imaginary numbers "real."

The discovery of special relativity was the first revolution. For anything that happens in the universe, relativity assigns to it four coordinates—three spatial ones telling us "where" it happened, and one temporal one telling us "when" it happened. The distance between two events is now measured in this four-dimensional spacetime. The mathematician Hermann Minkowski, a contemporary of Einstein (Einstein's teacher, in fact, who called Einstein a "lazy dog"), realized that rather than simply say that the equation for calculating distance was $x^2 + y^2 + z^2 - t^2$, because time is imaginary this distance becomes equal to $x^2 + y^2 + z^2 + (i \times t)^2$. Doing that square in the last bit restores the minus sign in front of t^2. (Some of you may have realized that t should be multiplied by the speed of light, c, but in the best tradition of theoretical physics, I am assuming that the speed of light is equal to 1.)

It is the imaginary $i \times t$ that represents the fourth dimension. This is not how Einstein thought about it. In fact, Einstein called Minkowski's construct a "mental masturbation" (if you are a physicist, that's a great way of insulting mathematicians). However, this particular masturbation was crucial for Einstein when working out general relativity (the best revenge a mathematician could have on a physicist).

This is interesting, because this kind of a distance is not really a distance! Or at least, not the kind of distance we are used to. If you go to any university, like Oxford, you'll see rectangular greens crisscrossed by paths, because it is less distance to walk the diagonal line than it is to walk along the edges of the green. Mathematically, this amounts to $AB + BC \geq AC$ (the sum of the first two sides is equal to the third side only when the triangle has an area of size zero). This inequality is known as the triangle inequality (for obvious reasons—see the picture), and it is violated in relativity. In other words, going directly from A to C could actually take longer than going via another point B.

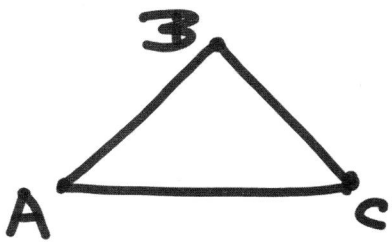

A standard inequality in Euclidean geometry, according to which the sum of any two sides is greater than the third side. It happens to be violated both in relativity and in quantum physics when interpreted accordingly.

An extreme example of this is when light goes from A to B and from B to C, with C ending up in the same spatial location as A (imagine a flash of light generated at A, traveling to a mirror at B, and bouncing back to return to A at a later time, which is the point C in spacetime). Then both AB and BC are zero (the paths that light takes are thus called null rays), while AC is some finite interval of time (however long it took light to make the round trip from A to B and back). The longest way round is, as far as relativity is concerned, the shortest way home.

From the fact that in the above example AC > AB + BC, all sorts of weird relativistic results follow. For instance, take the famous twin paradox we discussed earlier. The twin who travels away and comes back ends up being younger at the end of the journey than his stationary brother. The stationary brother ages by AC, which, as we saw, is greater than AB plus BC (the aging of the traveling twin).

We know that twins cannot travel at the speed of light, but the situational logic is the same for any speed of travel; it is just that the difference between the ages gets smaller the smaller the velocities involved are. (This is a shame, because if we could travel at the speed of light, we would not age at all!)

The time difference between things that travel at different speeds sounds like an esoteric phenomenon far removed from our ordinary lives. More like the stuff of science fiction. However, this is not true at all. Take two macroscopic crystals and heat them up to different temperatures. The fact that one is at a higher temperature is simply a statement that the atoms that compose it jiggle around more rapidly than the atoms of the crystal at the lower temperature. And that means the atoms of the two crystals are exactly like the twin brothers, traveling at different speeds and therefore aging differently. It also means that the frequency and color of the light emitted by the atoms of the two crystals are different. The light emitted by one cannot be absorbed by the other (if they are moving at the same speed, such as both being stationary, then the light emitted by one would be absorbed by the other). And this is a direct consequence of special relativity.

There is another way of making the crystals emit different frequencies of light. Suppose that we take a wheel spinning horizontally with respect to the ground and put the two crystals at different distances from the center of the wheel. Even though they are at the same temperature, they will now emit different colors of light, since each feels a different rate of acceleration (the greater the distance from the center, the bigger the acceleration). Different accelerations lead to different rates of time flow. Extending this to general relativity, this implies that time flows differently in different gravities—which is exactly why I observed earlier that gravity is just a manifestation of different flows of times in different spatial locations. How we understand acceleration and gravity will be important when we discuss whether the gravitational field should be underpinned by q-numbers and what experiments we can do to test whether gravity indeed is quantum.

The second revolution that made the imaginary numbers "real" was—big surprise—quantum physics. The key equation in quantum physics, the Schrödinger equation, is a so-called diffusion equation, but the diffusion of particles that it describes actually takes place in imaginary time. The square root of −1 is simply unavoidable in quantum physics. This has all sorts of interesting consequences, such as the fact that, unlike the generic diffusion equation, the Schrödinger equation is fully reversible. Otherwise, without the i, the equation would be irreversible and we would not have worried about the second law of thermodynamics! (Imaginary numbers are also unavoidable in the q-wave description of reality.)

Does the presence of imaginary numbers also mean that quantum physics violates the triangle inequality? The answer is yes—this violation is our old friend, Bell's inequality!

In quantum mechanics, the points A, B, and C represent measurements on quantum systems instead of points in spacetime. We can make measurements on a quantum system so close to one another that their outcomes are almost identical. In other words, the distance between these measurements, in the sense of the proximity of the outcomes, will be approximately zero.

Imagine that we conduct a sequence of such very close measurements: A is close to B, B is close to C, and so on. If we keep going like this, we will end up with a measurement, call it Z, which is very different from A, such that the distance between A and Z is equal to 1 (i.e., maximal). In other words, we add up lots of zeroes to get a distance equal to 1. The inequality AB + BC + ... > AZ is thus violated (because 0 < 1). Recent experiments I was involved with and that were implemented by Marco Genovese's group in Turin tested exactly this quantum property with photons. And if you perform this on two entangled particles, you will obtain Bell's inequality. (Incidentally, when one particle is involved, the inequality is sometimes called the Leggett-Garg inequality—or, as I like to call it, temporal entanglement.)

SPACETIME, CAUSALITY, AND THE QUANTUM

The whole seemingly paradoxical nature of quantum physics stems from the linearity of quantum operations. I won't go deep into linear algebra here, but for those who know it, this is the story: if we can execute two different unitary transformations on a given system, then we can also execute them in a conditional way by controlling which unitary transformation is applied depending on the value of a qubit ("unitary" just refers to a particular kind of transformation that is embedded in the Schrödinger equation). Specifically, if this additional qubit is in the state 0, we apply one of the two unitary transformations, whereas if the qubit is in the state 1, we apply the other one. The case of greatest interest to us is when the qubit is initially in a superposition of the states 0 and 1.

The upshot is that this process leads to entanglement between the qubit and the system undergoing conditionally the two unitary transformations, and the entangled system has undergone a unitary transformation that is a combination of the two different unitary transformations. This transformation is the key element in the Schrödinger's cat thought experiment as well as being at the root of quantum computation. One could say without much exaggeration that a controlled unitary operation is at the heart of every genuine quantum experiment (which, ultimately, usually involves the final measurement of the qubit after the controlled unitary has been applied, and this measurement too can be seen as a unitary transformation).*

---

* After almost a hundred years of quantum physics, we still find some consequences of the above transformation surprising. For instance, relativistic Lorentz transformations are represented by unitaries in quantum theory. This is natural, as the amplitude in quantum physics ought to be a relativistically invariant entity (strictly speaking, it is only the probability that needs to stay invariant, but this is not relevant for the present discussion).

Imagine that we prepare a spatial superposition of a particle. The qubit here is the spatial location of the particle, which provides the control. If the particle is in one of the two locations implied by the qubit, we apply what is known as a Lorentz boost, which basically means we impart some linear momentum to it; if the particle is in the other location, we do nothing. Now we have a superposition of two different velocities of one and the same particle.

According to special relativity, the clock of the particle in the moving branch of the superposition runs slower than the clock in the branch in which the particle is stationary. Now imagine that we interfere these two branches. This would result in the particle being localized at a single location in space (as caused by the interference) but in a superposition of two different *times* (i.e., a superposition of younger and older versions of one and the same particle). Quantumly, this entity is in a superposition of two proper times (that is, the times read off by each clock). As we said, if two things are possible classically (such as moving at two different speeds), then we can always bring them into simultaneous existence via quantum superposition. And we can make superpositions of superpositions and so on, ad infinitum.

Suppose now that we have two space-like events, which are events that cannot causally affect each other, as they are separated by a distance larger than the time it takes light to travel between them. For instance, two church bells striking 1:00 p.m., one in Paris and one in Stockholm (they are in the same time zone and I am assuming that the churches are perfectly synchronized), are events that are space-like separated, because it takes light about six milliseconds to travel between the two places. This means that there are observers who will think of these events as being at the same time (for instance, any Parisian who is stationary with respect to the churches), but there are also observers who will think that one of the bells rang before the other and vice versa (depending on the way the observer happens to be moving). When you mix in quantum physics, you can actually prepare a superposition of these—you could have a branch in which the observer perceives that the bell strikes first in Paris and then in Stockholm, and another branch in which the observer perceives these two events in the opposite order. This is something that is—in my view, mistakenly—called a state of "indefinite causal orders." It is more proper to say that each element of the superposition has a causal order, but that an observer outside of the superposition certainly has to acknowledge the superposition and cannot say more than that. This also means we have a superposition of different arrows of time. One observer will say that as time flows, it first encounters the chime of the Parisian church followed by the Stockholm one, while for the other observer time flows "backward" in the sense that the chime of the Stockholm church happens earlier than the Parisian one. All of this is a straightforward consequence of combining quantum physics with relativity.

One might get the impression that anything is possible in this world of quantum relativity. However, the exact opposite is closer to reality. The laws of quantum physics and relativity are very prohibitive and stringent (Einstein always emphasized that the more a law excludes, the better it is). Light always travels at the same speed for all observers, and no information can travel faster than it. All observers have to agree on the laws of physics; if one observer says that a photon entered one of your eyes (say, the left one), then all other observers (no matter how they are traveling) have to agree.

This is something I've always found surprising, but it does illustrate what I call the subtle tightness of the laws of physics. Add to the above two churches another one, this time in Rome. They all strike 1:00 p.m. at the same time for someone stationary with respect to them. Now, let's discuss the orders for other observers. You'd probably say that all orders must be possible. Well, let's see! If we abbreviate the Rome, then Paris, then Stockholm order as RPS, then we can describe six possible permutations (RPS, RSP, PRS, PSR, SPR, SRP), each corresponding to a different arrow of time. The interesting thing is this: according to relativity, not all permutations are possible. For simplicity, or perhaps because my geography has never been strong, I am assuming one spatial dimension, in the sense that the churches lie on the same line, but the result is true for events in the three-dimensional spatial world, too. There is a very simple reason for excluding all possible permutations of events. If there exists an observer that sees RPS and another one that sees RSP, then for an observer to see PRS, they would have to travel faster than the speed of light. Of course, relativity tells us that this is not allowed. These relativistic limitations automatically constrain quantum physics. Namely, we cannot prepare a superposition of these observers, either. The story would be different if the events were time-like, since then all observers would have to perceive the events in the same order.

Quantum physics is magic, but not in the sense that the impossible becomes possible. Quantum physics is magic, to paraphrase Arthur C. Clarke, because it permits a form of information-processing technology exponentially more advanced than anything allowed by the laws of classical physics. But it contains no paradoxes or inconsistencies, only things that might be counterintuitive to the uninitiated. And it does rule out a host of things that are otherwise—logically speaking—perfectly possible. Which brings us to our first speculation about the origin of quantum physics.

TOTALITARIANISM
VERSUS EGALITARIANISM

I'd wondered before about what Thales would make of the quantumness of the universe. Bryce DeWitt used the phrase "quantum totalitarianism" to illustrate the fact that anything that couples to a quantum system has itself got to be quantum. Not that quantum physics bosses other theories around, but because of the consistency of the overall description. Now I'd like to talk about another feature of quantum physics that I decided to call "quantum egalitarianism." You might think that egalitarianism is in inevitable conflict with totalitarianism. However, quite the opposite happens to be true, at least as far as quantum physics is concerned (some people say that in communism you can have both, but they are—I am quite certain—very much mistaken).

Egalitarianism emerges when we compare relativity and quantum physics. We saw that the theory of relativity unified space and time into one entity called spacetime. This might give us the wrong impression that time is just another dimension of space (the fourth). However, time is different, and even though relativity "mixes" space and time, when relativized they still maintain their separate characters in a few different ways. First, there is only one dimension of time, while there are three dimensions of space. Why this is so has given rise to countless speculations, each of which probably deserves a book in its own right. Second, in relativity, time, unlike space, comes with the imaginary unit! Third, things behave differently in space and in time. For instance, in classical physics (including special and general relativity):

(a) An object cannot be at two different places at any given instant of time.

(b) An object can be in the same place at two different instances of time.

(c) Two things cannot be at the same place at the same time.

(d) Two things can be in the same place at different times.

In quantum physics, both (a) and (c) are violated. Things can be in a superposition of different places at the same time (that's the whole point of quantum physics). Also, two (or more) quantum particles of the same type—say, photons—can occupy the same spatial location, even though all of their other properties (such as their polarization and frequency) might also be identical. Even different particles, such as a photon and an electron, can occupy the same point in space at the same instant of time. There is more. As we've seen in this chapter, an object in quantum physics can exist in a superposition of two different times while located at the same place. This actually makes the property of "being in a superposition" completely symmetric with respect to space and time. Superpositions apply to everything—they are egalitarian.

It's tempting to conclude that the superposition principle transcends, or perhaps gives rise to, space and time. This is an old speculation. For instance, physicist and philosopher Carl von Weizsäcker suggested that space is three-dimensional because the smallest quantum system, a qubit, requires three parameters to be fully specified. This is further underpinned by the mathematical property that relativistic transformations can be represented in the same way as qubit transformations. This could be a coincidence, but it is very suggestive of the fact that the qubit and space dimensionalities are not independent quantities.

Eddington's logic was even simpler, and it involved no quantum physics. He argued that any measurement is a comparison of two entities each having two limits. Take two rods, one of which is our standard for measurement. Since there are two ends to each rod, a measurement involves four entities. He then claimed that this leads to a four-dimensional spacetime that one can, according to Eddington, separate into a three-dimensional space and a one-dimensional time. Although Eddington's argument is not quantum, because all measurements must be subject to quantum physics, so too must Eddington's non-quantum view of spacetime arise out of measurements. This does not mean that quantum physics is prior to spacetime or is in any way "more fundamental." However, it is clear that quantum rules can be phrased independently of spacetime, while measuring distances and times cannot be done independently of quantum physics.

Given that quantum physics is sometimes seen as a "meta" theory (in Einstein-speak)—that is, quantum physics is a set of rules that has to be applied to whatever other classical theory you might have—this conclusion may not be surprising. After all, classical mechanics becomes quantum mechanics after the q-numbers and quantum rules have been incorporated. Classical electrodynamics becomes quantum electrodynamics, classical field theory becomes quantum field theory, and so on. Totalitarianism and egalitarianism strongly suggest that general relativity should become quantum general relativity— quantum gravity.

LOGIC IS DEFINED SO THAT MOTION IS SIMPLE

I've always been intrigued by Einstein's thoughts on physics and geometry. In one of his essays on this topic, he says that reality is defined by geometry plus the laws of physics taken together. Each of them can be tweaked individually, says Einstein, but so long as the other is adjusted in tandem, then the outcome will describe one and the same reality.

What he had in mind is as follows. Whether you think of the underlying spacetime as flat and the laws of motion saying that objects take curved (by gravity) paths, or you say that the spacetime is curved from the start and that objects follow straight lines (geodesics), the outcome will be one and the same reality regardless.

This is captured beautifully by John Archibald Wheeler's catchphrase "Time is defined so that motion looks simple." In other words, the way we think about time—as a one-dimensional, uniformly flowing stream—is actually arbitrary. The most commonly used notion of time is simply such because it makes the resulting laws of dynamics simpler to manipulate! We could always change how time is defined (for instance, clocks could tick at different rates at different points in space), but that would affect how we express the laws of motion.

And in some geometries, laws of motion do indeed look simpler than in others. In Newtonian three-dimensional Euclidian geometry, objects affected by gravity follow curved trajectories called parabolas (pretty complicated to describe mathematically). But in Einstein's general relativity, which is based on a four-dimensional curved manifold (a kind of mathematical object that corresponds to physical spacetime), objects follow straight lines (very simple in comparison with parabolas). All of this strongly suggests that time and space are arbitrary assignments and ought not to be considered as fundamental.

Now, I'd like to change Wheeler's dictum in order to apply the same reasoning to quantum physics. Here the relationship is not so much between geometry and physics, but instead between logic and physics. What do I mean by this? Well, classical Boolean logic and conventional quantum physics contradict each other. So in order to correctly interpret our experiments, we need to modify at least one or the other. If we hold Boolean logic fixed, we're forced to use a hidden-variable formulation of quantum physics, such as that of de Broglie or Bohm. Unfortunately, this makes the equations of motion very complicated. They are so cumbersome that even the proponents of this interpretation of quantum physics calculate things using the conventional Schrödinger equation and then "translate" the results back into de Broglie–Bohm mathematics (I kid you not). Alternatively, we could upgrade classical Boolean logic to quantum logic, in which case the laws of motion are indeed very simple (they are basically the same as Hamilton's classical equations, but with quantum numbers; in other words, "everything is a q-wave").

So there you have it. Reality is made up of different components, and our description of it necessarily invokes laws of physics, laws of logic, and laws of geometry (and perhaps even more). When we test our theories, we test the whole package;

we cannot say definitive things about any of them individually, and it is only about their totality that we can reach clear conclusions.

George Orwell said, "Freedom is the freedom to say that two plus two makes four. If that is granted, all else follows." Boy, was he wrong. In quantum physics, 2 + 2 makes $4 \times \sqrt{2}$ (remember the violation of Bell's inequality). Then the rest follows. Only in the classical world does 2 + 2 = 4, but our universe is not "made" that way.

You could say, in short, that our universe must, assuming we want to keep the laws of physics simple, be somehow both logically and geometrically curved. And that curvature brings us, inevitably, to the quantization of gravity.

Chapter 6

WHY QUANTIZE GRAVITY?

As soon as Heisenberg discovered quantum physics as it related to atomic physics, physicists wondered if the electromagnetic field ought also to be treated quantumly. As we've seen, Einstein had already argued for the existence of particles of light, photons, but his arguments were merely heuristic—he argued by analogy and didn't have a robust theory to back it up. Still, his analogy was ingenious: he noted that the formula for the entropy of a box containing a gas of atoms has the same mathematical expression as the formula for the entropy of the same box containing only radiation. This led him to speculate that radiation ought also to be composed of "atom-like" particles.

But Heisenberg had a better argument. He used the fact that constituents of matter, such as electrons and atoms—which were already established to be quantum—couple to the electromagnetic field. Because matter generates electromagnetic radiation and absorbs it, it would seem weird if matter obeyed one set of laws and radiation a different set of laws. In fact, Heisenberg's conclusion was even stronger than that. He argued that a quantum electron coupling to classical light would simply be inconsistent, and that it would violate his uncertainty relations.

The argument is a precursor to information-theoretic logic. An electron can interact with both the electric and the magnetic field components of the electromagnetic field. Heisenberg argued that the position of the electron could be measured by its effect on the electric field, while its velocity could be measured by its effect on the magnetic field. Since the electron obeys quantum physics, we know that its position and velocity cannot be measured simultaneously. But if the electric and the magnetic fields were just classical numbers—as the classical theory of electrodynamics says they are—then we could measure them instead to determine the electron's position and velocity exactly. A contradiction!

To maintain consistency, we need the electromagnetic field to be quantum mechanical, too—that is, we need to quantize it. This is a powerful argument, which of course is not a proof that things have to be this way. One can never prove mathematically that Nature ought to be in a certain way. For instance, it could be that there is something we don't know about that would prevent the electromagnetic field from coupling to electrons this way. It could also be that some kind of (purely classical) noise is preventing us from measuring the electric and magnetic fields at the same time. It could even be that during the experiment in which we imprint the position and velocity of the electron into the field, the electron itself ceases to behave quantum mechanically and becomes classical.

None of these other possibilities turns out to be true, but to rule them out one ultimately needs experimental evidence. And by now we have conclusive evidence that the electromagnetic field is quantum. We can in fact map out the exact state of the electromagnetic field in terms of photon numbers via tomography of light.

I'll describe one such experiment briefly. Imagine a tiny optical cavity between two mirrors within which we can trap light. The distance between the mirrors is comparable to the wavelength of light—typically we are talking about sizes on the order of a millionth of a meter. Now, one way to create a single photon in this cavity is to send an excited atom through the cavity, as the atom can be engineered to deposit a photon while between the two mirrors. (Experiments of this type led to a Nobel Prize for the French physicist Serge Haroche in 2012.) In a similar way, one can measure via tomography the general state of light in the cavity—that is, the number of photons—by sending a stream of atoms through.

Such experiments have enabled us to build a good understanding of the quantum properties of the electromagnetic field. But how—or, indeed, whether—gravity is quantum is still completely unclear. In fact, some people consider the question of whether gravity is quantum or not to be the greatest open problem in physics at present.

There are heuristic arguments for why gravity ought to be quantum. First of all, there is the Einstein-like argument by analogy. It turns out that when gravity is weak, its mathematical description is identical to Maxwell's description of electromagnetism. In other words, gravity behaves like a wave, much like light does. This was confirmed in 2016 through a direct detection of gravitational waves. The experiment was a tour de force. It took half a century to complete and resulted in the most sensitive and accurate measurements of vibrations in history. However, this was all about the existence of classical gravitational waves and doesn't tell us whether or not they are ultimately quantum. But a Heisenberg-style thought experiment could also be done with gravity. A gravitational field has electric-like and magnetic-like components. And, because matter couples to gravity, then if gravity was truly classical, one could violate the uncertainty principle as far as matter is concerned. So Heisenberg's argument leads to the same conclusion as it did earlier: the gravitational field should be quantum (at least the part of it that resembles the electromagnetic field).

A second argument for a quantum theory of gravity emerges from our desire to keep the principle of energy conservation intact, the importance of which we explored earlier. If particles interact at a distance, without any in-between mediation, then a problem with the conservation of energy arises. It's easiest to see it when we discuss a concrete two-body problem—say, involving the Earth and the Sun. What energy conservation means in that context is that, however the Earth and Sun move, the sum of the energies of the Earth and the Sun is the same number at all times.

Now, the Earth changes its speed as it moves about the Sun. The Earth's orbit is an ellipse along which the Earth travels at various speeds. For instance, when it's farther away from the Sun, it slows down. This means that it loses kinetic energy. However, we know that any information from the Earth takes about eight minutes to propagate to the Sun. So before this information about the reduction in Earth's kinetic energy arrives from the Earth, the Sun doesn't "know" that it needs to increase its own speed in order to "make sure" that the total energy stays the same. The question then is: if the Earth gave up some energy and the Sun has not yet absorbed it, where does this "excess" energy reside?

An argument due to Hendrik van Dam and Eugene Wigner says that if relativity is correct (namely, if there is no instantaneous action at a distance), then two particles distant from each other cannot interact if energy is to be conserved. Although one can redefine energy to allow for interactions, these kinds of redefinitions always have a non-local character; they invariably involve some retrocausal effect from the future on the past. And we've already seen that retrocausality is not a useful notion in physics. But here is where the notion of the field as the mediator of interactions bears fruit: the Earth imparts its energy loss on the gravitational field in its own vicinity, and this then propagates like a wave toward the Sun. At each instant in time, the interaction (the energy exchange) is always local in character, as energy exchanges from point to neighboring point.*

---

* This view also restores the idea that we call Newton's third law, aka the law of action and reaction. When a moving billiard ball hits a stationary billiard ball, the energy of the first ball transfers to the second, which then starts to move as the result of the collision. This is action and reaction. But the billiard balls do this by a direct-contact interaction. If, on the other hand, two things interact but never touch each other (like the Earth and the Sun), then without the gravitational field connecting them, Newton's third law is violated. This really doesn't make sense—either conservation of energy and Newton's third law are laws, or they're not. Newton, in his words, considered action at a distance to be "so great an Absurdity that I believe no Man who has in philosophical Matters a competent Faculty of thinking can ever fall into it." With the gravitational field present, on the other hand, Newton's third law is perfectly well restored. The "action and reaction" then happen first between the Earth and the neighboring field, then within the modes of the field, and finally between the field and the Sun.

So far, none of this involves quantum mechanics. So let's involve it! A mass such as the Earth (albeit smaller, to make the example more feasible without changing the power of the argument) can be in superpositions of two (or more) places at the same time. Not only that, but the energy of the mass could be different in each of the superposed positions. If we follow the above logic that things happen locally and are conserved locally, then the field in one of the branches ought to behave differently than the field in the other branch, since the mass in each branch of the superposition itself behaves differently. This is our old friend the Schrödinger's cat scenario in a different guise. It implies that the gravitational field needs to get entangled with the mass. Entanglement encapsulates this ability to respond to two things at the same time—namely, to the mass in two different places simultaneously. And since it can become entangled with the mass, it clearly has to have quantum mechanical features itself.

Notice that entanglement is invoked in order to preserve local conservation laws. This reinforces my argument that entanglement is local and not at all spooky. It is precisely *because* of entanglement that the world is *not* spooky! Entanglement saves us from having to add all sorts of weird things to physics in order to account for our observations. In the spirit of John Wheeler, you might say that entanglement is there to keep the universe simple.

Unfortunately, the entanglement between a single mass and the underlying gravitational field is impossible to detect conclusively (at present!), since we cannot directly measure the quantum degrees of freedom of the field. For that, we need another mass in a superposition that couples to the gravitational field. I will return to this idea later, after we've set the

A particle in a superposition of two locations acts on another particle to simultaneously attract it gravitationally to both locations. In quantum mechanics, the equivalence principle is obeyed in each branch of the superposition in which the upper particle exists.

scene properly. I will show you that a very generic argument can be made, underpinned by quantum information theory, that two masses are sufficient to witness quantum features of the gravitational field. This will lead us to a feasible experiment that could provide us with the first piece of evidence that Einstein was actually wrong! General relativity cannot remain classical and still account for all such experiments—it would have to be quantum.

That's not to say there are no difficulties in making gravity quantum—if there weren't, I suppose we'd be there already. However, the challenges are not what they are regularly claimed to be in popular accounts about why gravity is difficult to quantize. Because they come up repeatedly in the rest of this chapter, I want to summarize them here:

- **Entanglement.** Entanglement is not a problem. In fact, as we've seen, entanglement is part of the solution. If gravity is not quantum, matter has nothing to entangle itself with. This raises a host of issues like the apparent violation of energy and momentum conservation.

- **Equivalence principle.** The equivalence principle is not a problem. In the classical limit of quantized gravity, the equivalence principle of general relativity holds. Entanglement here happens to be crucial, too. It will make sure that the equivalence principle holds in all possible branches. A particle superposed in two places attracts another particle simultaneously to both places.

- **Time.** Time is not a problem. There is at least one possible way of thinking about time in quantum gravity, the Page-Wootters formalism, that appeals to a quantum physicist. Here different times are just seen as different quantum states of gravity entangled with the rest of the fields in the universe. But this is by no means the only way of thinking about time in quantum gravity. It is possible that time is not fundamental at all.

- **Measurement.** The measurement problem is not a problem. We've cleared this up earlier. Measurements are understood through entanglement. The same would be true in the quantum theory of gravity and is no different from quantum electrodynamics.

- **Non-local observables.** We are sometimes told that what is observable in quantum gravity cannot be located at a point. However, this is not true. All observables in quantum gravity can be defined locally at each point in space. The relevant field components in quantum gravity follow gauge invariance (just like the electromagnetic field). Don't worry if this sounds a bit technical at the moment.

- **"A field on itself."** Here there is an apparent problem of self-referentiality. All fields are phrased in terms of spacetime. For instance, the electromagnetic field can be described by all of its electric and magnetic components at each point in space and at every instant of time. But gravity is spacetime itself! So how do we talk about spacetime "sitting on top of itself"? This is more of a technical mathematical issue than anything to do with physics. I will show that it's not a problem that prevents us from quantizing gravity. The gist of the solution will be that spacetime is an arbitrary convention to label things and need not be fundamental.

- **Nonlinearities.** Nonlinearities are not a problem. Nonlinearities also arise in quantum electrodynamics, for instance, since a charge creates a field that in turn affects the charge. When it comes to quantizing spacetime, this is sometimes presented as follows: two spacetimes, each of which is a solution to Einstein's field equations, do *not* lead to another solution of Einstein's equations when they are superposed! This is because the gravitational field has energy, and this energy in turn acts as a source of the gravitational field. But, just as nonlinearity does not prevent us from finding quantum electrodynamics, none of this is an obstacle to quantizing gravity, either.

Armed with this, we are ready to talk about the killer experiment to disprove Einstein.

BMV: THE ULTIMATE GRAVITY MACHINE

I'd like to tell you about testing quantum gravity in the lab. It's a common misconception that testing quantum gravity can only be done with stupendously high energies and huge accelerators like the one at CERN. People think that because gravity is by far the weakest force, probing its possible quantum features would require an expensive CERN-style experiment. This, too, is not a problem: it is, in fact, possible to conduct a much cheaper laboratory test.

As we know, force is the key idea in classical physics. All particles move because they are acted on by forces. We can push an object from one place to another because the atoms of our hand exert a force on the atoms of that object. Still, classical physics does not offer a deeper reason for the existence of force. Why does the Earth move around the Sun? Because the gravitational pull of the Sun causes the Earth to move this way. How does gravity do that? The answer, my friends, is blowing in the wind. As far as Newtonian physics is concerned, that's just the way it is.

Classical physics acknowledges that there are different kinds of forces, such as the electrostatic force (by which two electrons repel each other), the magnetic force (which describes how magnets interact), the gravitational force (which is basically responsible for the large-scale dynamics in the universe), and so on. But the mechanism behind any of the forces was not known before quantum physics.

Don't get me wrong. There have been many proposals for how gravity works. Perhaps the most ingenious one was due to Georges-Louis Le Sage, who imagined that the universe is pervaded by invisible particles that travel in all possible directions until they encounter an object like the Sun, which either reflects or absorbs them. The Sun acts like a shield, and how many particles it prevents from reaching the Earth depends on the square of the distance (it's the effective angular area). If you work out the math of this picture, it turns out that more particles are pushing the Earth toward the Sun than away from it—and this imbalance is what we call the force of gravity.

This theory presents a mechanism that naturally accounts for the fact that gravity is a force that varies according to the square of the distance between objects (the same is true for

A depiction of an old theory of gravity trying to explain Newtonian action-at-a-distance theory by using particles moving randomly and pushing objects around.

electrostatic forces, a law discovered by Charles-Augustin de Coulomb). Unfortunately, the mechanism predicts some other effects that are not observed in reality. For instance, the Earth should have more gravitational particles hitting it from the front than the back, in the same way that you experience more raindrops on your face than on the back of your head when you are running in the rain. This would imply that there is a frictional force acting against the movement of the Earth that should ultimately slow the motion of our planet. No one has ever seen anything like that (which is just as well, since then we would gradually end up spiraling into the Sun).

There were other ideas around, but the most fruitful one turned out to be due to Faraday and (perfected by) Maxwell, who took classical forces as far as possible without the help of quantum physics. Their idea was that a charge generates a field around itself, which then propagates into the space around it. This propagating wave then acts on another charge some distance away. We can see these kinds of effects quite readily in our macro world—for example, with the maritime force. French sailors were warned a few centuries or so ago, well before any idea of fields in physics, about the behavior of two ships that are docked in a harbor parallel to each other. Surprisingly enough, they may start to approach each other, and may ultimately crash into each other, due to their small oscillations up and down in the water. The explanation of this effect, which is remarkably similar to how we describe forces between particles coupled through fields, lies in the fact that the waves that exist between the two ships (representing a field) actually cancel each other out, while the waves going away from the ships don't. This means that, due to the conservation of momentum, the ships recoil toward each other.

As we saw before, the key idea behind the electromagnetic field is to make apparently spooky or weird forces act in a causal way—namely, the electromagnetic waves propagate at the speed of light. The same is true for the gravitational field. So if the Sun vanished now, it would take about eight minutes for us to feel the (devastating) effect. What's more, the force lines of the field emanate from the charge isotropically (i.e., the same in all directions). There are therefore more lines closer to the charge than further away from it. It is this density of the lines of the field that is proportional to the strength of the force. And, just like the imaginary particles proposed to explain gravity, the density of the lines of force drops off with the square of the distance (since the same number of lines penetrates a larger and larger area as we move away from the charge).* Le Sage's idea was not completely crazy and actually contains some germs of truth.

Fast-forward to quantum mechanics. There are no forces; it's all about energy. Things move in such a way as to minimize energy. When you put a charge into the electromagnetic quantum vacuum, it creates photons. When you have two electrons, each creates its own cloud of photons. Each of the electrons (with its cloud of photons) then interacts with the other electron (with its cloud of photons), which increases the energy of the whole. To decrease the energy, the electrons have to move away from each other, and this is how quantum physics explains the repulsion between two like charges, while opposite charges would instead minimize energy by coming closer to each other.

Finally, we come to quantum gravity. Putting a mass into a superposition of two different locations would create two

---

*   This is the way in which Gauss phrased the Coulomb law.

different ripples in the gravitational field, one in each location. Now, if gravity was classical, we would either have one ripple or the other, but not both. Only if gravity is quantum could both ripples propagate at the same time.

Chiara Marletto and I—and, independently, my colleague Sougato Bose—had the following idea. What if we put an additional mass in a superposition of two different locations as well? If gravity is quantum, the two masses would become entangled. If not, they wouldn't.

Marletto and I have a general argument that makes very few assumptions, and all of them are satisfied by Einstein's general relativity: (1) the interactions between the two masses are local, and (2) they are capable of supporting the interoperability of information, which means that all channels of information with the same capacity are interchangeable (this information-theoretic idea has been made exact by Deutsch and Marletto in their work on the constructor theory of information). Our argument immediately implies that if the two masses become entangled through gravity in the experiment, then gravity must have some quantum components.

Here we have the principle of conservation of information at play. Suppose that you think of quantum and classical physics in terms of information theory. As we've seen, one of the key aspects of this is that quantum physics can encode much more information than classical physics can. In order to entangle two qubits of information, such as the two spatially superposed masses, we need a channel between them whose capacity is at least as big as that of a qubit. In the case of the two masses, that channel is gravity. That's why, if gravity was entirely classical, the qubits would not entangle, as the channel connecting them would not have enough capacity.

There is a race on to perform the Bose-Marletto-Vedral (BMV) experiment. I am aware of at least four groups of researchers (in addition to mine) who are competing to implement it. My bet is that the two masses would become entangled in the experiment, but I ought to be careful here, since there are some heavyweights betting against it (my colleague at Oxford and physics Nobel laureate Sir Roger Penrose, for instance). Anyway, time will tell.

We could also use the BMV experiment to do the same for electrical charges. Namely, we could entangle two of them via the intermediating electromagnetic field. This would be a proof that the electromagnetic field is quantum, but we already have many other bits of evidence that this is so (starting with blackbody radiation, the photoelectric effect, and the Compton

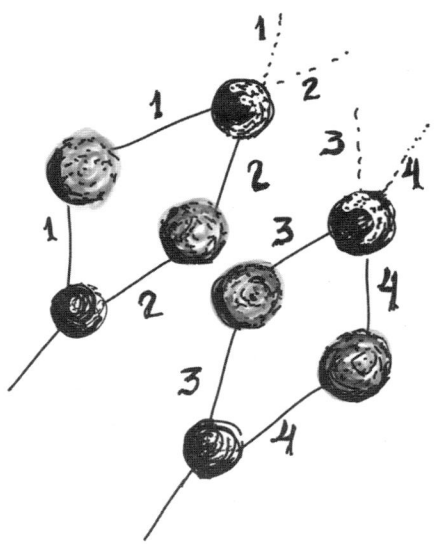

The Bose-Marletto-Vedral experimental proposal to test the quantum nature of the gravitational interaction. Two particles are put in a superposition of two different places each. The question is whether gravity can entangle them so that when the first particle takes the path 1, the second particle takes the path 3, and likewise for paths 2 and 4.

scattering, all of which were explained more than a hundred years ago). And, of course, we already know that the other two fundamental forces (weak and strong nuclear forces) are quantum, too, and the Standard Model has unified them with the electromagnetic force. The BMV argument could be used to test the quantum nature of any other new fundamental force, should such things be discovered in the future.*

WHAT COULD BE QUANTUM
IN QUANTUM GRAVITY?

Suppose that gravity can really generate entanglement between two superposed masses. Which of gravity's features would be characterized by q-numbers? For example, in quantum electrodynamics, the electric and magnetic fields are described by q-numbers. As we know by now, this means that they cannot be measured or specified simultaneously. The better we know the electric field, the less well we know the magnetic field. For gravity, the story could be similar. A gravitational field might be described as consisting of electric-like and magnetic-like components, one of which is good old Newtonian gravity, while the other component would be the gravity generated by the motion of masses (the gravitational analogue of Faraday's induction). These components could be characterized by q-numbers. If so, they would be subject to the Heisenberg uncertainty relations, just as position and velocity are. This would be sufficient to explain entanglement in the BMV experiment.

Other aspects of gravity could be quantum as well, and here the story gets even more interesting. If gravity just *is*

---

* The fifth force comes to mind here. It is frequently proposed to deal with the issue of dark energy.

spacetime, we would need to quantize the whole of space-time. One proposal, loop quantum gravity, does exactly that. Loop quantum gravity combines two different insights. One is by Roger Penrose on spin networks, and another is by Abhay Ashtekar, who thinks of gravity in terms of "loops." I don't want to go too much into detail here, but the upshot is that in loop quantum gravity, the lengths, areas, and volumes become q-numbers. It sounds weird, but that's quantum physics for you. In this framework, the better you are able to measure the side of a square, the less well you know its area! And with a cube, the better you measure its volume, the less well you know the area of its faces. The volume is like the energy of the gravitational field and the area is like the curvature of space. Therefore, the Heisenberg uncertainty principle would tell us that we could not simultaneously measure the energy of the field and the curvature of the space.

Other approaches to quantum gravity, such as string theory, also predict that gravity should induce entanglement between superposed masses. Fundamental objects in string theory are—as the name suggests—strings, which, unlike particles, have a spatial extent. In fact, the motivation for creating string theory was to avoid some uncomfortable mathematical implications that result from assuming that fundamental particles are point-like. String theory would also view gravity as a field, but one whose excitations are strings instead of particles. These strings would behave fully quantum mechanically.

The most straightforward approach to quantum gravity is so-called canonical quantization. This treats gravity like any other field, such as the electromagnetic field. In this theory, gravity would be mediated by a zero-mass particle called a graviton, similar to the photons that mediate the electromagnetic interactions. Gravitons, however, interact so weakly that an excited atom would take far longer than the age of the universe to emit one. This makes them very tricky to find. The Bose-Marletto-Vedral experiment was primarily proposed exactly to circumvent this issue.

So all the genuinely quantum approaches to gravity would agree that the BMV experiment ought to produce entanglement. In that sense, the BMV experiment would not be able to discriminate between string theory and loop quantum gravity. It would only be able to tell us that gravity is quantum as far as the relevant degrees of freedom are concerned.

Now, what happens in the (unlikely, in my view) event that the BMV experiment results in no entanglement? Of course, we are assuming that this outcome is not due to a mere engineering failure, but that the failure is instead fundamental: the experiment does not generate entanglement because gravity is not quantum. What then?

This state of affairs would be significantly more exciting for me (precisely *because* I am not betting on this outcome). We'd have to rethink quantum physics, since it would not be a truly universal theory. We would have an example of something in this universe—gravity—that is not quantum. It may not surprise you that, just as there are many versions of quantum gravity, there are also many theories out there that would be consistent with gravity not being quantum. Field theory in curved space is one example. In this so-called semi-classical theory, all fields other than gravity are quantum, and they sit on a classical spacetime background that includes gravity. The problem for this theory is to figure out how a classical spacetime background interacts with a massive object that is in a quantum superposition of two locations. Since here gravity is assumed to be classical and so cannot be superposed or entangled, the only logical conclusion is that the resulting gravity must come from one of the two positions, or some kind of statistical average of them. In general, for any semi-classical theory of quantum gravity, the gravitational field must act in such a way as to collapse superpositions in all other fields.

Related to this, Andrei Sakharov came up with what is known as induced gravity, which postulates that gravity is a consequence of the quantum effects in all other fields put together. In this picture, all the q-numbers in the universe give rise to classical gravity. Here gravity is not fundamental and so it is out of the question to quantize it.

Then there is the idea of entropic gravity, proposed by Ted Jacobson. It says that gravity is actually a thermodynamic effect of the other fields present in spacetime. This would make gravity a consequence of the first and second laws of thermodynamics. Therefore, this theory also predicts that the BMV test would fail to produce any entanglement via gravity.

There are many other possibilities, but the failure to see entanglement in the BMV experiment would definitely signal to us that we should rethink quantum physics, rather than force general relativity to have q-numbers.

THE WORST PREDICTION IN PHYSICS

One of the most amazing consequences of the quantization of the electromagnetic field is the appearance of the so-called zero-point energy of the quantum vacuum. This is a direct consequence, mathematically speaking, of the fact that in quantum mechanics, the electric and magnetic fields are characterized by q-numbers and so do not commute. The energy of the field equals the sum of the squares of the electric and the magnetic fields, but, because they are now q-numbers, even in the lowest energy state we cannot make this sum vanish. If we could, we would be able to specify the values of both the electric and the magnetic fields, which would violate the fact that they are q-numbers; in other words, if the energy of the quantum vacuum was zero, this fact alone would violate the Heisenberg uncertainty relations since it would mean that both the electric and the magnetic fields are simultaneously zero!

The existence of the quantum vacuum has some paradoxical consequences that we don't understand. The one I want to discuss here could signal trouble in the foundations of quantum physics.

To a good approximation, we can think of our universe as a box filled with the quantum vacuum. Each frequency that can occur has an energy equal to one-half of Planck's constant times the frequency.* In quantum electrodynamics, this enormous amount of energy is not a problem because only the energy *differences* are relevant. In fact, in many applications we can ignore the vacuum energy and set it equal to zero without any consequences, even though the vacuum state will still play a role. But there is one force of Nature that unfortunately ought to be affected by this vacuum energy in all its totality, not the mere differences between the vacuum energy and other energies. Yes, it's gravity.

According to general relativity, the *total* energy of the universe affects gravity. Even an otherwise empty universe would, according to quantum mechanics, produce an enormous

---

* The lowest frequency is inversely proportional to the size of the universe, and the highest is inversely proportional to the smallest possible length in the universe (which we'll assume is the so-called Planck length, though no one really knows).

amount of energy that general relativity tells us would gravitate strongly. In the equations of general relativity, this vacuum energy sometimes goes by the name of the cosmological constant.

Our best astronomical observations (to do with the rate of the expansion of the universe) tell us that the cosmological constant is tiny. It is the energy equivalent of having a single hydrogen atom in every cubic meter of the universe on average! However, if we use quantum mechanics to predict the quantum vacuum energy of the universe, it turns out to be 120 orders of magnitude bigger than what we see. Hence the worst prediction in the whole of physics.

There are a number of ways out of this problem. One is simply to say that our quantum calculation is not correct. We only added up the electromagnetic contribution (i.e., the contribution due to the bosonic fields), but if we were to add up all fields, various contributions might just cancel out (or almost cancel out). The logic here could be that the fermionic fields (e.g., electrons) produce the same vacuum energy as the bosonic fields but of the opposite sign.

Another way out is to deny the connection between the cosmological constant and the vacuum energy. Perhaps the cosmological constant is due to something else, such as the total mass in the universe (or maybe the fifth force).

Yet another way of resolving the apparent paradox is to deny the reality of this quantum vacuum energy, or to state that it has no gravitational influence whatsoever. Some physics heavyweights like Wolfgang Pauli and Julian Schwinger thought this way.

On the other hand, maybe this discrepancy tells us that there is a fundamental problem with quantum mechanics itself. Perhaps just as classical physics predicted that an infinite amount of energy should radiate from a blackbody, the huge amount of energy that quantum physics predicts in the vacuum* could also mean the breakdown of quantum physics.

So could the worst prediction in physics give us a clue about how to "fix" quantum physics? Should we, for example, be upgrading q-numbers to some other entities—say, the w(eird)-numbers? Perhaps, but given the success that quantum physics has enjoyed in the microscopic world of atoms, photons, subatomic particles, and molecules, it is difficult to see how one ought to modify it while preserving all the good explanations it has provided us with so far.

My personal feeling is that it might not be quantum physics that is at fault here. It could be that we should be looking into modifying general relativity instead. One can only speculate on how gravity ought to be modified. Sakharov's idea of induced gravity, which says that gravity is not really a fundamental force but is the result of all the quantum vacuum fluctuations, could fit the bill, but it has not really been developed properly. It might also fail to comply with the BMV experiment.

There is another remote prospect to consider. Perhaps we need new mathematics altogether to make progress in physics. Investigating this takes us to the place where the conflict between quantum physics and gravity is most acute.

BLACK HOLES

Perhaps the biggest challenge for quantum physics comes from strong gravitational fields. This is because when objects are

---

* Incidentally, this would also be infinite if there was no smallest length, but we don't currently know whether or not there is such a thing.

really massive and confined to a small volume, general relativity tells us that they become black holes. The gravitational pull of black holes is so strong that not even light can escape them, hence the name, coined by John Wheeler.

A black hole is truly a marvel. Much as the Sun forces the planets to revolve around it, a black hole can trap even light into orbiting around it. This sounds amazing. If you stood on the surface of a black hole (this would be hard to execute in practice, as you would need to engineer counterforces to prevent you from falling in and being squashed and elongated), you would actually be able to see the back of your own head! The reason is that the light emitted from the back of your head would orbit around the black hole and eventually arrive at your eyes.

Black holes appear to be a problem for quantum physics because their dynamics is irreversible as far as general relativity is concerned. Things that go into black holes do not come out. However, to reiterate, quantum physics is a fully reversible theory. Any dynamics going in one direction can always be—at least in principle—reversed. Some researchers therefore argue that quantum physics and general relativity are incompatible and that it is not possible to quantize gravity.

The issue was highlighted by Stephen Hawking and is called the information loss paradox: in quantum physics—in *any* reversible theory—distinguishable states remain distinguishable. An apple is different from a pear, and this remains so forever. In classical general relativity, when an apple and a pear fall into a black hole, their differences get obliterated by the strong gravitational field. Both become indistinguishable parts of the structure of the black hole. They seem to lose their individual identities. Inside the black hole, there simply is no way to discriminate apples from pears, or anything from anything else.

This is how things stand if general relativity remains a classical theory. But once we quantize it, maybe black holes are no longer as irreversible as they currently seem to be. Consider that, according to classical electrodynamics, atoms are not possible, since accelerated electrons would lose energy by radiating and would therefore fall into the nucleus. Atoms are not stable in classical physics, but quantum theory eventually came to their rescue. It could be the same story with black holes. Their classical properties would likely change when we quantize them.

Of course, we don't really know which of their properties ought to be quantized. We already have some indications that in a semi-classical theory, black holes actually radiate—this is called Hawking radiation. But even this picture, where space-time is just a classical background and other fields behave against this background, is not sufficient to quantize gravity.

Some have argued that if black holes are treated fully quantumly, then there is an apparent violation of the principle of monogamy of quantum entanglement (a phrase due to one of the pioneers of quantum information, Charles Bennett). This principle, which follows directly from the laws of quantum information, says that if two quantum systems are maximally entangled, they cannot be entangled with anything else. Why would radiating black holes violate this? Because if black holes radiated quantum mechanically, then this radiation would cause the inside and outside of the black hole to become entangled. This is all fine, but what happens when the inside is maximally entangled with the outside? It looks as though any further radiation would have to entangle with something else, but this seems impossible according to the principle of monogamy of entanglement.

This apparent paradox actually has nothing to do with black holes per se, but rather with how quantum physics explains radiation and evaporation more generally. It is fundamentally all about the fact that quantum physics is a fully reversible theory. Any process that takes place in one direction should be able to be reversed—so it should be possible to entangle and then disentangle a set of systems. Therefore, the issue really has to do with thermodynamics and quantum mechanics.

Let's take a few steps back to explain what's at stake here. Every object that absorbs radiation must eventually emit it: if you absorb energy, you get hotter, and things hotter than their environment tend to cool down. This cooling-down process is caused by the object radiating some of the energy back. An atom in its lowest energy state can absorb a photon, which gives it the energy needed for it to jump to a more excited state. When the atom later emits a photon back into the environment, it loses this energy and "falls" back to the ground state. In a perfect equilibrium, a body absorbs and emits the same amount of energy.

Let's think about this from an information-theoretic perspective. Initially, the energy sits in the electromagnetic field, manifested as a photon. There is no entanglement between the field and the atom, which is sitting in its lowest energy state. Because both systems are in definite states, each has zero entropy. The entropy of the combined system, then, is also zero. However, at the next instance, the photon is in a superposition of being absorbed and not being absorbed by the atom. The atom is correspondingly in its excited and ground states simultaneously. This is an overall entangled state between the atom and the field. They are maximally entangled when the state in which the atom is excited and the field has no photon is as likely as the state in which the photon has not been absorbed and the atom is in the ground state. Furthermore, the atom and the field are now each in a state of maximum entropy. However, the whole state still has zero entropy. What happens as the dynamics of our atom-field system continues? Well, the atom and the field now start to become less and less entangled until the atom returns to the ground state and the photon is emitted back into the field. The state has come back to where we started. Exactly the same analysis applies to our Alice-Bob-poison-cat experiment.

The black hole and its surroundings are in an analogous situation. The same "life cycle" could hold for any black hole: it first gets maximally entangled to its environment and then continues to radiate until becoming completely disentangled and vanishing. This would be how black holes evaporate, if this is indeed what they do. We have no experimental evidence of this fact, and it's extremely unlikely that this evidence will be available in the foreseeable future.

Despite our inability to wait for a long enough time to see if black holes evaporate, the main point is that there is no need to violate either quantum physics or general relativity to accommodate the dynamics of black holes. It is questionable whether we will ever be able to perform quantum experiments with— or within—black holes. But at present, there seem to be no a priori problems with having a strongly gravitating object that is also described by q-numbers.

Our current understanding of black holes also points to the interesting possibility that space and time are themselves discretized—that is, formulated such that they only take on values from a discrete set of possibilities. This would corroborate the intuition behind loop quantum gravity. The possibility may arise from a black hole's entropy.

The entropy of any object is the logarithm of the number of relevant degrees of freedom, such as the number of possible states that the object can occupy. In the case of the black hole, we can think of the number of units of its surface area that are the size of the Planck length squared (where the Planck length is 1.6×10^{-35} meters).

If we think of the Planck area as being the smallest area that can hold a single bit of information, then the black hole entropy is simply the total number of bits it can hold. In other words, take the area of a black hole and divide it by the Planck area to get the number of bits that the black hole can hold. It would certainly be fascinating if space and time were quantum numbers and those numbers came in finite units. But such units expressed, say in terms of the Planck area, are far too small for us to be able to detect, even indirectly.

There is a well-known result, called the Bekenstein bound, that says that a finite region of spacetime must also have a finite entropy. If we think of the black hole entropy as the number of possible ways in which spacetime can exist quantum mechanically, this suggests that spacetime must be discretized. I wrote about this at length in *Decoding Reality*, but for our purposes here, we are indeed very far from being able to prove anything of this kind. Discretization of spacetime would definitely help us because the current way we do quantum field theory leads to a number of infinities that don't make physical sense.* In this way, rather than being paradoxical, understanding black holes through quantum gravity could in fact improve the state of affairs in other applications of quantum field theory.

BLACK HOLES AS A TESTING GROUND
FOR QUANTIZING SPACE AND TIME

Because black holes involve such a huge gravitational field packed into a quantum-sized region of space, we have no choice but to take both general relativity and quantum mechanics into account when describing them. To make matters worse, because space and time are allowed to be arbitrarily small, the equations of general relativity produce infinities and thus break down at the center of a black hole (the so-called singularity).

However, there are now glimmers of hope. Recently I have been developing an idea that might get us somewhere by making quantum mechanics more like general relativity. With the help of some experiments, it could lead us to the center of our black hole, and to a unified theory at last.

Instead of taking ideas from general relativity and molding them to fit into quantum mechanics, I think we need to change the way we think about time in quantum mechanics. To make quantum mechanics more compatible with general relativity, we should try to treat space and time in the quantum world as we do when it comes to spacetime.

As we've seen, in relativity, time is interwoven with space in such a way that they are on equal footing. Like space, time can

---

* The throwing away of infinities in quantum field theory, a process known as renormalization, is one of its least appealing features.

bend and stretch, depending on the speed or strength of gravity involved. This is a far cry from the way most physicists talk
about time in quantum mechanics, where it is a fixed, external
entity—a series of steps through which interactions evolve and
nothing more. In traditional quantum mechanics, time is only
inferred from looking at other observable quantities, such as
the hands of a clock.

This inference of time is reflected in the way we tackle
equations in quantum mechanics—for example, how we
describe the position in space of a qubit, which can exist in a
superposition of several states at once.

A little over a decade ago, I began to consider an alternative
to the Schrödinger equation, which is normally what we would
use to describe the trajectory of a qubit over time. I read about
some experiments that showed that quantum entanglement
can exist not just between objects in *space* but also between
the states of the same object at two different *times*. This means
that particles sometimes seem to be affected by events that take
place in their future. This strange idea caused me to wonder:
if particles can be entangled over time, then perhaps time in
quantum mechanics isn't the steady, external ticking clock
we'd thought it was.

Along with my colleagues Joe Fitzsimons at the Singapore
University of Technology and Jonathan Jones at the University of Oxford, I have developed an alternative mathematical
approach.* Instead of describing the behavior of a quantum
object in space at one specific time, we describe the behavior
of quantum objects across all of spacetime at once. For example, when calculating how a particle should interact, we would

---

* See Joseph F. Fitzsimons, Jonathan A. Jones, and Vlatko Vedral, "Quantum
Correlations Which Imply Causation," *Scientific Reports* 5 (2016): 18281.

write down all the possible outcomes at all the possible points in space and at every instant of time (unlike with the standard Schrödinger equation, whereby only one instant of time goes into the calculation). The fundamental dynamics is the same— particles are still governed by the three fundamental forces. But instead of describing a particle's position over space at one point in time, we describe it over spacetime.

Just as time is relative in general relativity—meaning it varies from observer to observer—in our new description of quantum mechanics, time is no longer a fixed, separate entity. In practice, this means that when you are looking at two different states of a particle's position across spacetime, it isn't always definitive which came before the other. This ambiguity doesn't exist in standard quantum mechanics, where all observers agree on the order of events.

In our description, spacetime is an entangled web of quantum-correlated events all given in advance. Our description of quantum mechanical time matches up with the general relativistic notion of time, and so it could be that our novel way of doing quantum mechanics offers a portal into the next theory that will unify the two deepest theories in physics.

Over the ten years since we began this journey, my colleagues and I have started to test our theory. We have performed experiments with two types of qubits to illustrate that our approach works.

In one experiment, Marco Genovese and his group at the National Metrology Institute of Italy looked at photons and their polarization, which is the way that the electric field oscillates as a photon moves through space. In another experiment, published in February 2021, Jones used a quantum property of particles called nuclear spin to illustrate the same idea. The crux in both experiments is to measure the qubits at multiple times and calculate the quantum spin we are probing (whether it's polarization or nuclear spin) across space *and* time. Those measurements are then compared with what our theory predicts.

It turns out that anything that can be calculated using the standard approach to quantum physics can also be calculated using our spacetime version. Experiments have confirmed that our approach is a perfectly valid alternative to that of standard quantum mechanics. But because it shares a description of time with general relativity, our version is a concrete step toward a unifying theory.

Let's return to the center of a black hole. As we've seen, general relativity ceases to work when it encounters a singularity. But when we apply our new way of thinking about spacetime in quantum mechanics, we can incorporate an extra degree of fuzziness that will save us from these singularities. Just as there is uncertainty between a particle's position and momentum in quantum theory, there may similarly be an uncertainty between space and time. If this is true, it would mean that at the center of a black hole, regions of spacetime would be so small that space and time could no longer be discriminated from each other.

In principle, we could prove that experimentally. The test would be similar to how we check the uncertainty relations between the position and momentum of a single particle, only now we'd be testing the uncertainty between space and time. In more concrete terms, the more precisely we measure the spatial distance between two events, the less precisely we measure the events' temporal separation. We don't know if this is how Nature really works, but our approach could certainly handle it. Current experiments don't have enough resolution to test space and time at these minute scales, but the requisite technology is progressing rapidly.

Since a singularity is traditionally understood to be the point at which both space and time contract to zero, a quantum uncertainty in spacetime would prevent singularities from being possible—like I said, if our theory turns out to be right, space and time would no longer be distinguishable at the center of a black hole. Phrased another way, our theory says that space and time cannot possibly both have the precise value of zero simultaneously.

There is another surprising implication of our theory: quantum time travel. Closed time-like loops are loops in which spacetime curves in on itself, thereby establishing a route to the past. Although general relativity seems to allow for them, traditional quantum mechanics rules them out. Our version of quantum mechanics, though, seems to allow for them!

Think of an entangled pair of particles, particle A and particle B, connected through some interaction such that the measurement of one immediately affects the other. If particle A goes through a time-like loop, then this trip would create two copies of particle A, one younger and one older. Each of the two copies is expected to be maximally entangled with particle B, which did not go through the loop. Traditional quantum physics rules this scenario out as impossible. But if we use our theory to describe the situation, then the younger and older versions of particle A are simply entangled in time.

If our version of quantum theory is correct, then not only might we finally complete our imagined journey into the heart of a black hole, but we just might discover a route back in time, too.

We have journeyed to environments in which gravity is so strong and quantum effects are so severe that neither could be ignored. But could it be that surprises lie not in this direction, but rather in a totally different domain? Unlike black holes in galaxies far, far away, the domain I have in mind is right under our noses. In fact, it includes our noses.

Chapter 7

ENTANGLED LIVING SYSTEMS

One of the ideas that immediately occurred to the early practitioners of quantum mechanics was that, since quantum physics was able to explain how atoms bond into molecules, perhaps quantum effects might also be relevant in the biological domain. After all, biological systems are all about using chemistry in order to generate the energy needed to sustain themselves in a state far from equilibrium.*

However, there were views to the contrary as well—that quantum mechanics is either irrelevant or detrimental to life. Physicists like Wigner thought that life collapses quantum superpositions and forces quantum processes to become

---

* Being in equilibrium with your inanimate surroundings simply means that you yourself are inanimate, too (i.e., dead).

classical. But he also thought that the laws of quantum mechanics made it exceedingly unlikely that replicating entities would ever arise in the fully quantum universe.

There were even more complicated views, such as those of Niels Bohr. In his lectures titled "Light and Life," he applied his concept of complementarity to chemistry and biology. Living organisms have intricate structures that function at the atomic level (e.g., sense organs capable of detecting just a few photons). Therefore, Bohr claimed that a detailed analysis of their behavior must depend on the theory of atomic physics. However, a new aspect of complementarity, in addition to the original Heisenberg relations, arises here: the use of these methods in studying life creates a paradox. By conducting physical experiments to probe the atomic processes involved in biological functions, according to Bohr, the experimenter inevitably disrupts the living state, effectively destroying the vital activity under investigation. And it is not just Bohr who held these views. He influenced a number of physicists, primarily Heisenberg, Pascual Jordan, and Pauli, who all subscribed to a similar philosophy. In fact, they frequently ventured beyond biology and saw complementarity at work even in human perception. Jordan, for instance, was fond of commenting that introspection, through which we the observers observe ourselves, must be limited in a manner similar to how the position and the velocity of a particle are.

He expanded on Bohr's ideas about psychology. Bohr had identified a strong parallel between the challenges of measurement or definition in atomic physics and the paradox of self-analysis in psychology. As William James and others had observed, self-analysis creates confusion between "I" and "me," or the roles of observer and observed. In Bohr's framework, this translates to an uncontrollable disturbance caused by the analyzer's interaction with the subject under examination. Jordan maintained that this perspective sheds new light on the classic problem of self-analysis: the freedom of individual action. Philosophers had long debated this issue, dividing them into opposing camps of free-will advocates and determinists, assuming that only one could be correct. However, according to Jordan's reading of Bohr's ideas, complementary analysis reveals that both concepts—free will and strict determinism— are necessary to account for human experience and they are not in contradiction. We perceive free will when considering future actions and apply causal reasoning to past events. In essence, we believe we are free until the moment we recognize that we have already chosen. Free will and determinism are therefore complementary, in Jordan's take, in line with the uncertainty principle.

Of course, such broad speculations quickly got out of hand, frequently leading to pseudoscience and confusion. Pauli, a person who was famous for his no-nonsense approach to physics, became a patient of Jung's and tried to understand bits of Bohr's philosophy in the realm of psychoanalysis. It is amazing that the person who at the age of nineteen had the audacity to get up after Einstein's lecture and say "You know, what Einstein has just said isn't so stupid" was the same person who suggested that the psychological phenomenon of split personality could be an instance of the complementarity principle applied to human neurology!

Undeniably, some aspects of perception are quantum. For instance, there are experiments with human vision that test a very simple property—namely, how dim a light our eyes can detect. What's interesting here is that each subject of the test reports an incredibly consistent average value of the intensity of light they can see. Thus, contrary to what we might think, the subjectivity of different individuals is not relevant here.

To be sure, in each run of the experiment, the same flash is produced, which is sometimes reported as seen and sometimes not. But this randomness is not dependent on the individual subjects. All of this is explained via the quantum nature of light. In the experiments on human vision, each pulse of light contains an indeterminate number of photons. This is why the subjects sometimes detect three photons in the pulse and sometimes detect ten. But the human threshold is on average seven photons. Anything above this will generate the neurological impulse necessary for us to become aware of the light pulse. This is why the same subject, when targeted with the same intensity of pulse, does not always say, "Yes, I see something." It is because the light pulse is made up of photons.

Now, why seven? The reason lies in the so-called dark counts. Human receptors of light are subject to random—so-called thermal—fluctuations. That means that even when there is no light around, the receptors will fire at a certain rate due to being at a high temperature, which itself provides energy to spontaneously produce neurological signals. Therefore, our eyes will mistakenly tell us that we are seeing light even when there is no light around. We know this because even in the dark, when we close our eyes, we occasionally see flashes.

Now here comes a cool punch line. The rate of this spontaneous firing is the energy equivalent of six photons! So our eyes evolved to have a threshold above this number in order to make sure that we are seeing not just random thermal fluctuations, but rather that we are seeing something that is really out there.

One wonders if the same can be said of our hearing. How many phonons (the particles of vibration) do we need to hear a sound? I haven't been able to find any research on this, but it may well be that our hearing is not as refined as our sight—which is, after all, our dominant sense—and that the quantum nature of sound may not play an important role in hearing.*

Even among the early speculations of pioneers of quantum physics, there are many valid points raised and many issues to which we still have no satisfactory answers. It is, after all, perfectly legitimate to ask how much biology is reducible to chemistry and, ultimately, to quantum physics. But as I have argued throughout this book, if modern physics tells us anything, it is that there is no special place for observers. Interestingly, Jordan, the person to name the field of quantum biology, in which scientists study the role of quantum physics in living systems, came very close to denying any difference between living and non-living observers. His view was that biological systems are just amplifiers of the underlying microscopic random quantum events, and they are no different from other, non-biological measurement processes.

Here is how the science historian Richard Beyler sees this aspect of Jordan's thinking (most of Jordan's writing is in German and hasn't been translated):

> Despite the claims, amplifier theory had delivered nothing holistic or teleological. Jordan's own explanations said as much. Geiger counters and Wilson cloud chambers, he wrote, worked according to the same principle as biological amplifier processes. An unpredictable

---

* I half humorously wrote a paper titled "Can We Hear the Sound of a Quantum Superposition?" It is an interesting topic well worth exploring.

microphysical event set off an avalanche-like process that, itself, ran quite deterministically. The result was a macrophysical manifestation of the micro event: say, an audible click representing the ionization of an atom by an X-ray photon. Biological amplification was one member of this category of measurement processes that produced objective "tracks" by inducing a "decision" from the microphysical entity under observation. The amplification process thus constituted an incursion into the totality of superposed states of the microphysical center that set off the process. From a biologist's perspective, there was nothing about the mode of action of a Geiger counter or a cloud chamber that suggested the stability and apparent self-maintaining qualities of a living organism.

These thoughts tie in perfectly with our discussions. Different observers (whether conscious or otherwise) are really just different ways of amplifying quantum events. While one qubit entangled with another is sufficient for the purpose of quantum measurement, amplification simply entails *more* qubits joining the entanglement. And it is this amplification that makes things harder to undo.

There are many other interesting questions at the interface of quantum physics and biology. There is the possibility that life itself owes its existence to quantum physics. Here I have in mind the fact that even small-scale quantum computations could be much more efficient than their classical counterparts. Perhaps the evolutionary "search" for the first replicators was indeed quantum mechanical in nature. The related question is whether biological systems exploit quantum phenomena such as the existence of large superpositions and many-body entanglement.

Ultimately, the first significant step would be to try to entangle living systems and show that there is no complementarity between being alive and being quantum. Both can exist and be confirmed simultaneously. The reason I think this is an important portal into the future of physics is that maybe, just maybe, entangling living systems is actually impossible. This might indicate that life needs classical information and cannot function in the presence of quantum information, an outcome that would go directly against the principle of the universality of quantum information that I argued for earlier (that quantum information-carrying media should "win out" over classical information-carrying media upon interaction between them).

JIM MARTIN

There is an important aspect of science that is frequently neglected in popular writing—the funding. My adventures into the world of quantum biology have largely been possible because of the generous support of private individuals and foundations. None of it has really come from governmental grants. Just like in the (good old?) days of Victorian science.

James Martin was possibly the biggest donor ever to Oxford University. He bequeathed £200 million in 2010 for an institute whose purpose would be to work on the most challenging problems facing humanity in the twenty-first century. The resulting Oxford Martin School was then set up in central

Oxford—close to the Bodleian Library—to house the researchers working on the problems Martin thought would define the next one hundred years.

Luckily for me, Jim Martin liked my research and was the first supporter of my adventures into the quantum world of bacteria. I think Jim would have enjoyed these results very much had he not passed away in 2013. He kept asking all sorts of questions about the consequences of quantum physics for our understanding of the macro world. What were the implications of an entirely quantum mechanical universe? That's what his vision was all about.

BUCKYBALLS

I'd like to tell you about a very famous breakthrough experiment in physics that was carried out almost thirty years ago. There is a research group in Vienna, led by the recently minted Nobel laureate Anton Zeilinger, that sought to experiment with a very large molecule, fullerene, which contains sixty carbon atoms. Why fullerene? Because it is relatively easy to handle in the lab—it has a high degree of symmetry (it looks very much like a mini version of a soccer ball). Synthesizing this organic molecule earned someone a Nobel Prize in chemistry, and the molecule itself is affectionately known as the buckyball. Both this nickname and its chemical name were inspired by the American architect Buckminster Fuller, whose structures had a high degree of geometrical symmetry, much like the molecule itself.

The quantum wavelength, the degree to which we should expect an object to behave like a q-wave, is inversely proportional to the mass of that object. In this case, the wavelength of the buckyball is a thousandth of a billionth of a meter—tiny indeed. If an object is much larger than its quantum wave-length, then that object behaves classically for all practical purposes. In the case of the buckyball, its classical quality is a thousand times bigger than its quantum wavelength.

Zeilinger's group had six theoretical physicists visiting at the time of the experiment, all trying to predict what would happen if the buckyball was sent through two splits. Given what we said about the smallness of its quantum wavelength, none of them predicted that the buckyball would be in a super-position of going through both slits at the same time. Each of the visiting theoreticians offered different arguments as to why one would never see any quantum mechanical features in a molecule like this. (Just so that you know: I wasn't one of the six physicists, but I'm sure I would have made exactly the same mistake.)

Fortunately, the theoretical naysayers didn't put off Zeilinger and his colleagues—they still went ahead and did the experiment. To make matters worse (for quantum mechanics), these molecules are initially deposited on a surface and, in order to be detached from it and launched toward the slits, they need to be heated up to about a thousand degrees Kelvin (much hotter than room temperature). At such a high tempera-ture, the buckyballs emit lots of radiation on their way to the slits. During each molecule's flight to the slits, it emits on aver-age six photons of light.

Every time we fire a molecule, we ask, "Does it go through one of the slits at a time? Or can it be superposed and go through both slits simultaneously?" The results of the exper-iment were conclusive: each molecule did go through both slits. The experimental results they got from diffracting the buckyballs are indistinguishable from what you would get by doing the same experiment with photons!

Now, why doesn't the buckyballs' emission of photons tell us anything about which slit the molecule is going through? After all, if each molecule emits six photons on average during the run, then even a single one of these photons should be sufficient to tell us the location of the molecule. In other words, the emission of photons should constitute a bona fide quantum measurement, as the molecule gives away the information about its location. This kind of logic was also used by some of the visiting theoreticians to argue that the buckyball would never be able to interfere. The reason this logic is faulty—and

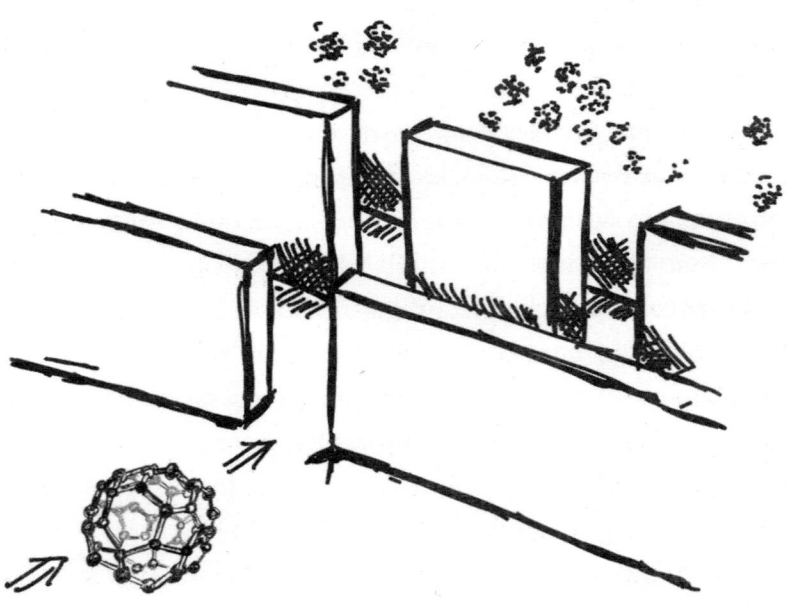

One of the groundbreaking quantum interference experiments of the 1990s. Markus Arndt and his collaborators managed to send one fullerene molecule through two slits at the same time and demonstrate that it can exist in a superposition of two places at the same time.

we know it is, because the experiments had succeeded—is both subtle and simple to understand. The wavelength of the emitted photons is much larger than the distance between the slits (that distance being about a millionth of a meter). So this is not enough to tell us anything about the exact location of the molecules. The photons tell us that the molecule is somewhere in the slit region, but we cannot say anything more precisely than that. So even though buckyballs are large molecules, emit lots of photons, and "should" be classical, it turns out that they behave fully quantum mechanically.

In the language that I like, the buckyballs do get entangled with their environments by emitting photons, but this entanglement doesn't prevent them from behaving like waves because it doesn't contain the information regarding which slit they go through. In this sense, the photons are a bit like the piece of paper in Alice's experiment that Bob uses to answer her question about whether he sees a definitive state of the cat. That piece of paper doesn't entangle with the relevant degrees of freedom of the experiment, and it was important that this was so, for otherwise it would not be intact at the end of the experiment.

The question remains: How far can we go? Can objects bigger than buckyballs also exhibit quantum behavior? I have been arguing that this is one of the most exciting and crucial questions for the whole of physics, perhaps even for the whole of science. Does quantum physics hold at every scale? We don't know, which is why we need to probe different regimes and knock on different doors.

Some people are proposing to do a double slit experiment with a virus—which is even bigger than a buckyball—because, as we have seen, some people say that something odd might happen with quantum mechanical weirdness when applied to living systems. I think we're relatively close to superposing a virus in two different places. So might even a virus go through two slits? There are some smart people who have different answers to this question, which is why it's an exciting issue.

For instance, there's the famous Ghirardi-Rimini-Weber (GRW) model, which simply says, "Look, it's 'obvious' that large objects are not in two places at the same time." GRW would argue that we've got to modify quantum mechanics, since somewhere along the way of increasing complexity it'll collapse, although no one yet knows where. So the GRW model simply changes the Schrödinger equation by adding a term to it that causes a collapse of quantum superpositions.

This kind of modification is problematic for a number of reasons. One is that collapse occurs in a specific observable, usually the position of a particle. It forces the position to become a c-number, but it leaves other quantum features as q-numbers. Second, it breaks Bohr's correspondence principle, which was the cornerstone of Heisenberg's quantization. This principle says that we should leave the classical equations of motion intact and only convert the c-numbers into q-numbers. GRW violates this because it forces even the classical dynamics to change and become irreversible. This has another knock-on effect: since the evolution becomes irreversible, there ought to be an entropy increase (heat generation) associated with it. Nothing like that has so far been observed.

QUANTUM LIFE

Maybe some macroscopic quantum effects already are out there in Nature. This could mean that we might not have to work that hard at creating them artificially. If Nature took four billion years to evolve some kind of small-scale quantum computers through natural selection, why not reverse-engineer these natural processes and adapt them to our purposes?

Already there is evidence for entanglement in complex many-body systems. A couple of decades ago, an experiment was performed on a large (for quantum purposes) chunk of salt (a few milligrams' worth). They tested the response of atoms inside the salt to an external magnetic field. Next they tried to simulate the experimental results first using classical physics, then using a half-classical and half-quantum simulation (excluding entanglement), and finally using a fully quantum mechanical description that included entanglement. They found that you have to take into account the entire entangled state to correctly describe their experiments. Therefore, this is a clear instance of a macroscopic entangled state involving nearly a million billion billion atoms.

I have to be a little bit careful because we physicists tend to think about relationships between natural sciences the way Lord Rutherford did. He uttered one of my favorite statements of all time: "Science is either physics or stamp collecting." If you look at the laws of evolution, they're not the same kind of laws as Newton's laws of motion. Evolution does not have the same power to predict. We don't have the same explanatory power, either. I think it is beyond any doubt that evolution captures key aspects of biology, but I bet that we'll be able to reduce it to something like the laws of quantum physics. It's a bit ironic that Rutherford himself was a Nobel laureate in chemistry; apparently he kept the Nobel Prize in his bathroom, and legend has it that he was hiding it there because he was ashamed of having won the prize in chemistry rather than physics.

For every physicist like Rutherford, there is another physicist who thinks that all sciences can never be reduced to physics. Another Nobel laureate, Philip Anderson, published a very influential paper called "More Is Different" in the journal *Science* in 1972. There are no equations in this two-page paper. Anyone can read it, and I highly encourage you to do so. Anderson said that there are complex systems whose behavior can never be reduced to the underlying laws of quantum mechanics, even in principle. Personally, I would be very disappointed if the world turned out to be like that—if there is indeed a disconnect between the domains of the small and the large. But I recognize that the world doesn't care if I'm disappointed or not. The universe is indifferent to our human desires.

Then there is Schrödinger, who comes in and says that maybe life requires some new laws of physics. I think that it's not an accident that the person who thought very deeply about entanglement also thought about scaling this effect up to chemistry and biology. His famous book *What Is Life?* was read by many physicists and convinced them to go into biology. His book was basically a call to arms saying that biology is far too important to be left to biologists. This is true even now. Many major breakthroughs in biology, such as uncovering the structure of DNA, were made using techniques from physics. Today, physicists hope to design some kind of novel quantum spectroscopy that will help biologists and chemists better understand the complex structures of molecules and their dynamics.

Maybe life could happen without the universe being quantum mechanical. I don't know, and neither does anybody else. This is why these kinds of experiments are one of the paths to new physics.

PLANTS AND BIRDS

Perhaps the best evidence we have for the existence of quantum effects in biology comes from photosynthesis. Graham Fleming's group at Berkeley has been focused on this for a long time, and they finally reached a breakthrough in 2006.

The entire photosynthetic process is extremely inefficient. From a plant's initial capture of light to the transfer of this excitation to the reaction center and finally to the chemical processing of this energy, the overall efficiency is somewhere between 1 and 5 percent. However, one step in the photosynthetic process—the transfer of captured photons into the reaction center—is nearly 100 percent efficient.

In this step, roughly speaking, the photon is transferred from molecule to molecule until it reaches the reaction center. The efficiency of this step is so high that it almost violates the second law of thermodynamics. For a long time, biologists couldn't explain it. After all, biologists usually think in terms of classical physics. But eventually a few quantum chemists from Berkeley said, "Maybe *quantum* physics holds the solution here."

The idea is that as the photon's energy cascades from molecule to molecule until it reaches the reaction center, it actually takes all possible routes simultaneously by way of quantum superposition. As far as we know, this is the only way of explaining how this process takes place with almost 100 percent efficiency. To be sure, researchers are still debating whether this is actually the solution.

Classical physics also can explain high efficiencies if waves are used instead of random classical motion. Random classical hopping from molecule to molecule is a bit like a drunkard walking home from the pub—they are as likely to take a step in the right direction as to take a step in the opposite direction.

This leads to a smaller probability of making it home (or, in the case of photosynthesis, making it to the reaction center). The key question is therefore whether classical deterministic waves or quantum deterministic waves are at play here.

The most extensively studied living systems are the purple bacteria, which usually live one thousand meters below sea level. There they receive between three and five photons per hour on average. Unless they are very efficient at capturing these photons, they wouldn't have enough energy to survive in such extreme conditions. The circumstantial evidence from experiments so far points to the fact that these bacteria, too, exploit quantum mechanics in order to capture photons with extreme efficiency, though we certainly need to perform many more experiments in this direction to reach a conclusive answer.

Magnetoreception, the ability to detect the Earth's magnetic field, offers yet another example of quantum effects in living systems. We already know that we can influence chemistry in the lab by changing the external magnetic field: by changing the orientation of the field, we can alter which chemical reactions take place. But, of course, simple inanimate chemistry is not the same as living systems. How do animals navigate in the face of the Earth's magnetic field?

Over the course of a year, the European robin flies from northern Europe to the equator and back again. A research group in Frankfurt led by Wolfgang Wiltschko and Roswitha Wiltschko captures robins somewhere along these flights and subjects them to experiments in artificially generated external magnetic fields. They observe the direction in which the robins move in order to test how and if the robins manage to orient themselves relative to the magnetic fields. The idea is that robins must be using some kind of internal compass during their great migration flights.

What they found in 1971 was unexpected. When they put the magnetic field in one direction, robins flew along that direction, but when they completely reversed the direction of the field, the robins did not reverse course. In other words, the robins did not notice if the magnetic field was reoriented in the Wiltschkos' setup.

If we exchanged the north and south of a magnetic field, the needle of a compass would rotate 180 degrees to align itself with the new north. We know that lots of other animals use a compass made of some kind of magnetic substance, and when the Earth's magnetic field changes, so do the magnetic substances of the animals' compasses. This can all be explained classically. But the robins don't do that. This killed the idea that robins use a simple classical compass. The Wiltschkos' subsequent experiments solidified the fact that there isn't a simple classical explanation in the case of robins. They discovered that when one changes the angle of the external magnetic field, the robins do actually change the direction of movement. So robins can't tell the polarity (they can't tell when we flip north and south), but they can tell the inclination of the field. The question is how they can do it.

Then came a proposal by Klaus Schulten and refined by Thorsten Ritz. The idea was that robins exploit the quantum dynamics of two electrons inside a molecule that's sitting in the retina of the bird's eye. The proposal effectively says, "Even biological systems like European robins use a quantum chemical mechanism in order to tell the direction of the magnetic field, though not its polarity."

I did some research on this, because I hypothesized that robins also harnessed entanglement, the signature effect in quantum mechanics. This was something that, at that time, was considered highly unlikely because we are talking about a macroscopic and living system. With a few colleagues of mine from Oxford and Singapore, we showed that indeed the electrons in Ritz's model were entangled. The resulting paper appeared in one of the premier international journals, *Physical Review Letters*. This was also the only paper of mine in my scientific career to have been reported in the *Daily Mail*! (I probably should have retired there and then.)

Our most surprising discovery was that the entanglement between electrons in robins seems to last a long time, about a tenth of a millisecond. This may sound very short, but by the standards of electronic entanglement it is in fact a very long time. At the time of our discovery, this was much longer than anything that could be done artificially with electrons in physics labs.

It really surprised me that Nature's large objects at high temperatures (the faster movement of particles at higher temperatures produces a great deal of thermal noise, which can disrupt entanglement and more generally disrupt quantum effects) could better exploit quantum mechanics than we can with our quantum technology in much friendlier settings! Maybe we will one day reverse-engineer what robins do and use their "technology" as a building block for quantum computers.

BACTERIA ENTANGLED WITH LIGHT

Some time ago, the journal *Small* published a paper entitled "Biophotonics: A Nanophotonic Structure Containing Living

Photosynthetic Bacteria." It's the culmination of my work within the Oxford Martin Programme on Bio-Inspired Quantum Technologies, and it's the first experiment of this kind that involves entanglement and a bacterium that is provably alive.

As is the case with all good ideas, the conversation about this experiment started in a pub (the Royal Oak in Oxford, since you've asked). Dave Coles (the first author of the paper and the person who did the experiment) and I were sitting in that pub one Friday evening in 2015. Coles was telling me about his experiments on green sulfur bacteria's light-harvesting antennae. These large organelles, called chlorosomes, are responsible for capturing light for photosynthesis in these bacteria, which can live deep down in the Black Sea. At about one thousand meters below the sea surface, they get no more than a few hundred photons per second on average. To put that into perspective, it would take them almost half a quadrillion years (about thirty thousand times the age of the universe) to gather the equivalent energy contained in a Mars bar, so they had better be good at catching photons!

In a previous experiment, Coles had already shown that he could strongly entangle light with chlorosomes on their own. This experiment made a splash and was published in *Nature Communications* as "Strong Coupling Between the Chlorosome of Photosynthetic Bacteria and a Confined Optical Cavity Mode." Its main selling point was that it showed that large biological systems such as chlorosomes can be in a genuine quantum mechanical state together with light.

During that evening at the Royal Oak, I told Dave that he could do even better. Why not take a living bacterium and try to entangle *it* with light? Each green bacterium contains about 250 chlorosomes. So if each chlorosome can be in an entangled state, could we therefore get an entire living system to become quantum entangled with light? This seemed like the trailer for a science fiction movie! Could something be fully quantum and alive at the same time?

Dave and I kept discussing what I thought were crazier and crazier ideas, but—to my surprise—at each point Dave would confidently say, "Yes, I think I can do that, too." It was thrilling. Needless to say, the pub closed at a critical juncture in our discussion, so we had to improvise. We got some beer and wine from a local off-license shop and went to my house.

We stayed up until four in the morning. I must admit, I don't remember much of the discussion (read: at some point I become tired and emotional), but I just knew that it was momentous. The next day, I flew to Singapore (to visit the Centre for Quantum Technologies), but Dave went to the lab to conduct preliminary experiments. And they worked! He sent me an email with images showing that living bacteria could, in fact, be entangled with light. I was very excited (I remember that I could not sleep that night in Singapore, and I don't think it was the jet lag). Dave wrote up a paper based on the results that later became the cover story in the journal *Small*.

The way the experiment shows that light and bacteria are entangled is by detecting a change in the emitted light compared with when they are not entangled. When the relevant q-numbers characterizing the energy of the bacterium interact with the relevant energy q-numbers of light, then the joint energy changes and is no longer the sum of the individual energies. This is pretty much the same mechanism as when atoms bond together to form molecules; it's just that in this case we have a "molecule" composed of the bacterium and quantum light.

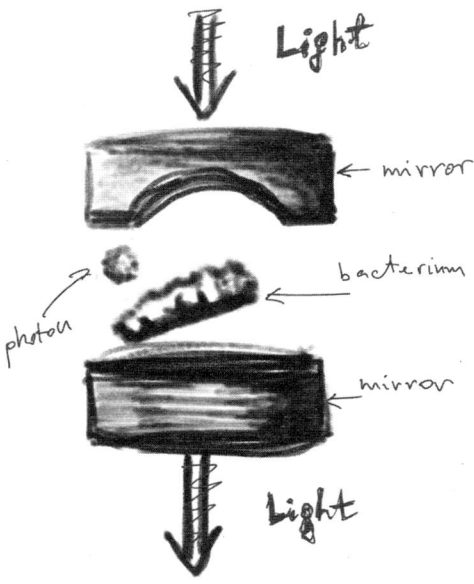

An artist's impression of the optical setup where a living bacterium was trapped between two mirrors to create an entangled state with light.

But here's why I think these kinds of experiments could open another portal into new physics. Coles's ideas can be used to put bacteria in a state in which they are dead and alive at the same time. This would be hugely exciting, as it would be an almost real instance of Schrödinger's visionary thought experiment. However, as the next step toward that, I thought of realizing a hybrid system made of a qubit and a much larger living microorganism—a tardigrade!

ENTANGLED QUBITS AND TARDIGRADES

Some creatures never die. I recently read an article about a bunch of tardigrades (also known as slow steppers or water bears) in which the author described how tardigrades were sent to outer space to test their robustness. During most of their journey, the conditions were highly inhospitable: there is no air out there, and the temperature is less than three degrees Kelvin (about one-hundredth the temperature in your living room). But they returned to Earth intact. Not only that, but there were more of them upon return than upon take-off! (They had actually managed to multiply in these adverse conditions.)

Tardigrades have evolved an ingenious strategy in which they can "switch on" a state of completely suspended animation. In this state, formally known as a "tun," their bodies dry out and appear as lifeless balls, and their metabolism plummets to as little as a hundredth of 1 percent of its normal rate. Tardigrades can survive as tuns for decades because their energy consumption is so minimal. (Anyone interested in combating global warming might want to take note.)

The incredible resilience of tardigrades gave a couple of colleagues of mine a mind-blowing idea. Rainer Dumke of the Centre for Quantum Technologies in Singapore and Tomek Paterek of the University of Gdansk had already collected the Ig Nobel Prize (a spoof of the Nobel Prize, based on a play on words) in biology a couple of years earlier for studying the magnetic response of cockroaches. That work was within classical physics. Now they thought of doing something even more exciting: coupling tardigrades to quantum computers (this is what physicists think of as "exciting").

For the past five years or so, Dumke has been developing a superconducting qubit-based approach to quantum technologies. In order to make his qubits stable, he needs to cool his superconductors down to temperatures that are a thousandth of the temperature in outer space and a ten-thousandth of the temperature in your living room. Could tardigrades survive this interaction with supercool qubits?

The answer turns out to be yes. And not only do they survive low temperatures, but they also survive entanglement with superconducting qubits, which results in a lower energy state for the joint system than either could achieve on its own. This lower energy is precisely what Rainer's group measured, and it provides the main witness of entanglement. Again, it's the interaction between the relevant q-numbers that leads to lower energy. But they went much further—they managed to prepare arbitrary superpositions of this hybrid qubit as well as entangle it with yet another superconducting qubit. This is a very exciting result for physics, as it pushes quantum mechanics even further into the macro domain, exactly what's needed to realize Schrödinger's thought experiment. However, it is also interesting to a biologist, such as our collaborator Nadja Møbjerg of the University of Copenhagen, as Dumke's experiments demonstrate that tardigrades can survive record-breaking low temperatures and pressures.

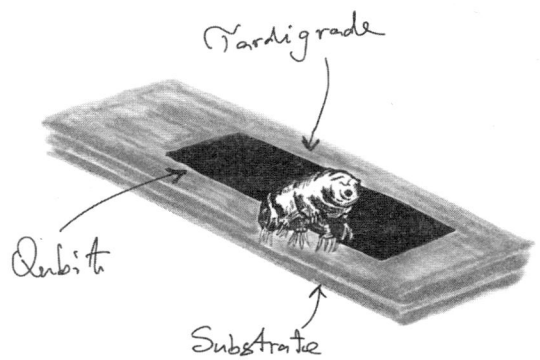

A sketch of the experiment performed by Rainer Dumke in which a living tardigrade was integrated with a quantum superconducting bit.

About a thousand tardigrades have been stuck on the Moon since 2019 after an Israeli rocket carrying them crash-landed there. Most scientists believe that the tardigrades are still alive and kicking (in case you asked, we named the tardigrade that survived Dumke's experiment Neil Wormstrong). Does quantum physics play any significant role in how they are able to survive extreme conditions? And could their phenomenal survivalist makeup give us a clue about how to colonize other planets? While it may take us some time to understand all these big questions, one thing is clear to me: I feel privileged and lucky to be part of this wonderful collaboration and hope that more is soon to come.

All of this ties back to our "observing the observer" experiments: we could contemplate how one could get a simple living system to confirm that it is in a definitive state even though it is really in a superposition of different states. That's always the game in quantum mechanics experiments I have in mind. As we said, the difficulty with biologically simple systems like viruses, bacteria, or even tardigrades is to get them to confirm that they are in a superposition of distinct states. They can't speak to us, but they can communicate with their actions. For instance, the bacterium in a cavity could be simultaneously able and not able to synthesize. This means that it is able to absorb and not able to absorb an incoming photon. It is this last feature that would constitute a proof of being in a definitive albeit entangled state. However, even this would be a challenging experiment, although simpler than the one involving Alice and Bob. Luckily for me, the Gordon and Betty Moore Foundation is funding me on the way to this adventurous portal.

AI AND PHYSICS

To realize Schrödinger-like entanglements, do we really need to have actual living systems, or could we have something that simulates a living system? One natural candidate that already exists is a computer running artificial intelligence software. This software would clearly be able to assess whether it is observing a definite outcome or not. But, in my view, it is not equivalent to having a fully conscious entity involved in an entangled state.

The term *artificial intelligence* is a misnomer. Yes, it is artificial if you think the things we produce are artificial (which itself is debatable). However, it is not intelligent by any reasonable definition. AI is simply a sophisticated way of doing spectacularly efficient searches while optimizing for various criteria.

Take chess engines, for instance. They are now far superior to any human player. They follow simple algorithms in which different chess pieces are assigned different values (pawn = 1, knight = 3, queen = 9, and so on), and then the algorithm investigates as many possible moves as time (relative to the computer clock speed) allows. Each investigated combination is assigned the sum total of all values and the highest value then provides the route followed in the game.

As mind-numbingly stupid as this strategy is, thanks to the fantastical speed of computers it is enough to beat the best player in the world, Magnus Carlsen, every day of the week and twice on Sundays.

Now get this: every AI algorithm is a variant of the chess engine algorithm. None of them has what one might call an intuition. They can come up with pleasant-sounding music or poetically concocted prose, but it is clear that none of them can discover a new law of physics. Forget physics—they actually struggle with recognizing patterns that my two-year-old daughter recognizes with her eyes closed and her hands behind her back.

The reason is that we don't have the foggiest idea about how the human mind works. We do not even understand what kind of computation is going on behind our own eyes. And there, I said it—computation. Some people even doubt whether human consciousness is based on computation. Maybe none of the computational paradigms we've so far discovered fit the purpose of mimicking our brain.

There is an idea called the Church-Turing-Deutsch principle that basically says that physics equals computation. More elaborately, the principle asserts that any physical process can be faithfully simulated (so that it is indistinguishable from the original) on a universal computer. If we therefore think that what's going on in the brain is just physics (and what else could it be?), then it ought to be possible to make a computer behave exactly as a brain does.

But (there is always a *but*) we now know that quantum computers can do exponentially many more things than classical ones. It is therefore also possible that while classical computers could never be conscious, a quantum computer could. For all that we know, a quantum computer with only a thousand qubits—meaning that it could manipulate 2^{1000} states all at the same time, more than there are atoms in the universe—could be programmed to be conscious. But we'd still first need to understand how the human brain actually works.

In any case, to those of you worried about the Terminator scenario, worry not. The bottleneck of the whole problem is understanding how neurons in the brain interact and work together to produce the magic we call human thinking. One of the most exciting questions is whether quantum effects play an important role in this. But who says that quantum physics is the last word as far as the laws of Nature go? It is possible that even quantum computers are not able to be conscious in the way that we are, but that a new theory of physics finally unlocks the door to the human mind. Call me old-fashioned, but I think we should devote all of our best efforts to finding this out.

Going back to the main issue of an AI being entangled with a quantum system, the immediate challenge is to understand the minimal number of qubits within the AI that would allow verification of both the entanglement and the definitive observation. Unlike the hard question of consciousness, this one is well within current technological reach.

Chapter 8

DETECTING
QUANTUM GHOSTS

Ghosts have fascinated us and haunted our minds from time immemorial—ephemeral entities that exist in a limbo between the living and the dead, entities that we can never see completely but which can affect us and terrify us. In physics, there are ghosts too. They're not scary or supernatural, but, as in traditional folklore, they stand for something we do not fully grasp. In the context of physics, a ghost is an entity, an element of reality, that we have to introduce into our theories in order to make them internally consistent. However—and this justifies the name—ghosts are claimed not to be directly observable. It will become clear what this means, but one example of ghosts goes under the name of virtual particles.

These particles, which cannot be detected, are the ones claimed to be responsible for the static forces in quantum field theory. So two electrons repel each other by exchanging the undetectable ghost photons. Needless to say, I argue that this is not the correct way to understand things and that we can in fact test for alternatives. Welcome to the fifth portal.

So far we've looked at quantum physics in the macro domain. In that domain, the tensions are strong between quantum and gravity and also between quantum and living systems. But even in the micro domain, where gravity and life are largely or entirely irrelevant, it could be that quantum physics needs to be modified or replaced by a deeper theory.

This chapter is the most challenging, as you've most likely not encountered these kinds of discussions before. It's to do with the very foundations of reality in the form of quantum waves. But the battle must be fought on all fronts. And this includes intellectual battles with one own's colleagues.

Every time we scientists submit a paper for publication, there is a so-called peer review process. Our colleagues are asked to read our submitted paper and evaluate it. This affects whether and how the paper needs to be revised and whether or not it gets published. I am saying all this not to then go on to complain about referees and the peer review process (I could, but it's boring) but because of an exchange I had with one of my recent referees.

The crux of the argument is about what constitutes the primary objects in quantum physics. What entities should we start with and build the rest of the world from? As you know by now, I have been advocating that we should start with q-waves and then derive everything else from them—particles, matter, light, and so on.

The referee, on the other hand, emphasized the fact that we ultimately know everything through our senses (aided by sophisticated experimental equipment, granted). It is therefore the process of detection that—according to this logic—ought to be primary, and that everything else ought to be derived from that. In other words, clicks in detectors are what the universe is made of. What is a photon? A photon is a particle that makes a click in the photodetector! What is an electron? A particle that makes a click in the "electrodetector." And so on . . .

DOES A PHOTON MAKE A SOUND WHEN THERE IS NO ONE THERE TO HEAR IT?

There is something intuitive about the referee's perspective, since, after all, science starts with a sequence of observations ("clicks" in our brain) that we then try to analyze, understand, and ultimately synthesize into scientific theories. However, our interpretation of observations does not come from nowhere but is instead based on the prior beliefs, views, and theories we hold even before we make our observations. In other words, there is no such thing as a pure observation, no such thing as a "naked" click. For instance, in order to make a photodetector, one has to already understand some rudimentary electrodynamics. It is for this reason that I think it is impossible to build physics on the assumption that clicks are the primary elements. In fact, even a single detection, a single click, is an extraordinarily complicated phenomenon.

If everything is ultimately made up of q-waves, why do we get clicks at all?

Let's take a rather simple experiment. We excite an atom, which emits a photon, which travels to our detector, which, upon absorption, makes a click sound. How are we to make sense of this in terms of q-waves?

Even a simple sequence of events such as this one is actually quite complex when we try to explain it in terms of the bare fundamentals. For starters, what is an atom and how did we excite it in the first place? Why did it subsequently emit a photon? Why did that photon travel through space to us and get sucked into the detector? How does the detector interact with the photon to register its presence? How did the detector produce a sound, and how did that sound propagate through the air and reach our ears? How did our brains then convert that into the sensation of a click sound?

I'd like to outline the best picture of the above experiment we have. Bear in mind that we have no clue how to explain human perception. "Why does a click feel like a click?" is a question no neuroscientist or philosopher can answer.

For starters, an atom is composed of more fundamental particles: electrons, quarks, photons, and gluons (plus a few other things that are not worth mentioning). Quarks are glued together into protons and neutrons via particles called—what else?—gluons. Protons and neutrons are also glued together into atomic nuclei. Electrons orbit these nuclei. For our purposes, nuclei are stable enough that we can talk about them without having to always refer to their constituents.

At room temperature, electrons exist in their lowest orbits around the nuclei. However, if we shine some light onto an atom, an electron from the lowest orbit will absorb energy from the light and "rise" to a higher orbit. Because the higher orbits are not stable, electrons tend to return to the ground state. As they do so, they emit photons.

These photons can generally emit from atoms in all directions. However, in one of these directions (or a small range of them), the photon will encounter a photodetector (in other directions, it will just continue until it encounters another piece of matter that can absorb it).

A photodetector is typically a conductor of electricity, but it only conducts when a photon hits it and excites one electron. This electron then gets accelerated by an electric potential and goes on to hit atoms in the detector and knock out even more electrons. Ultimately, the size of the current (i.e., the number of moving electrons) becomes large enough that the detector is able to notice it. What does that mean?

It means that in a normal state, the detector is an insulator and generates no signals (which is how it is calibrated). But when photons strike it, the detector gains enough energy to start to conduce a current whose moving electrons are easily detected. When this detection occurs, a sound is generated. This is the click we hear.

The click is a sound wave emitted by the detector that travels in all directions, which is how it eventually encounters our ear. The internal machinery of our ear then starts to vibrate, which generates a small current that flows through our neurons and into the depths of our brain. There the sensation of the click is finally generated.

So a click is an immensely complex process whereby an optical wave (laser) is converted into an electronic wave (an excited atom), which then emits another optical wave (a photon). This wave then generates *another* electronic wave in the photodetector in the form of a current, which ultimately results in a sound wave (vibrations of atoms in the air between the detector and us). The sound wave gets converted into a matter (mechanical) wave in our ear and again into an electron wave that travels down our neurons into the brain. This chain of events ultimately leads to the brain conjuring up the sensation of a click sound.

All in all, it's clear that a click is an "emergent" phenomenon rather than something fundamental. As recalled by James Boswell, Dr. Johnson's biographer: "After we came out of the church, we stood talking for some time together of Bishop Berkeley's ingenious sophistry to prove the non-existence of matter, and that everything in the universe is merely ideal. I observed, that though we are satisfied his doctrine is not true, it is impossible to refute it. I never shall forget the alacrity with which Johnson answered, striking his foot with mighty force against a large stone, till he rebounded from it, 'I refute it thus.'"

So if any components of a field—electromagnetic, gravitational, or any other—can "kick back" when you "kick" them, then they should be thought of as real. The footnote I'd add to Dr. Johnson's criterion is that if they can kick back at a quantum system, then they ought to be quantum themselves!

Bottom line: clicks are not the fundamental elements of reality. Q-numbers are, as they do the kicking.

Maybe the next physical theory will also contain the explanation of why a click sounds to us like a click (for example, how it is that we "hear clicks" while bats actually "see clicks" instead). One thing I am confident of is this: the explanation can only come from physics and not from chemistry, biology, psychology, medicine, philosophy, or mathematics alone.

But the bigger problem inherent in my referee's comment is that the referee is ultimately making the classical world the fundamental world. If any bits of our theory remain classical, we are in danger of having a half-quantum, half-classical description of reality. As I have been at pains to point out, all such descriptions suffer from inconsistencies. At least some of gravity's aspects have to be quantum. Observers also have to be quantum and, in fact, are no different from any other quantum system. Living systems, too, require us to use q-numbers in order to account for some of the behavior we have observed so far (and more will come in the near future, I am sure).

Surprisingly, one of the domains of physics that is still haunted by classical-quantum hybrids is electrodynamics, and more generally a class of interactions in physics known as the static interactions. Our current understanding of them is problematic, but this can give us a glimpse into new and deeper physics. At the root of the problem is a phenomenon known as ghost modes, which are—still—treated classically rather than quantumly. What would investigating the quantum reality of the ghost modes tell us? That even more things behave quantumly than we had assumed. And this is the way to go in order to push the boundaries of physics.

WHO YOU GONNA CALL?
GAUGE BUSTERS

Consider two electrons. How do they experience what is known as electrostatic repulsion? We've talked about this before—there is an electromagnetic field everywhere, and each electron disturbs it locally. This disturbance is propagated by the field from each electron to the others.

In the conventional formulation of this process, we don't need to think of the disturbances as quantum. Instead, the disturbances can be thought of as changes to the energy of the electromagnetic field that in turn are expressed as c-numbers. The electromagnetic field can be thought of as being made up of an infinite number of simple harmonic oscillators, and the energy of each of these oscillators changes when we introduce electrons into the field. This change in energy is positive when the charges are alike and negative when the charges are opposite. To minimize energy, the charges repel each other in the former case and attract each other in the latter case.

The problem with treating these electrostatic interactions as classical changes of energy is that electrons are (of course) quantum mechanical and can be in two or more places at the same time. How do we describe a charge's disturbance to the electromagnetic field when the charge is in a superposition of locations? By now, we're used to the logic of the situation—remember Schrödinger's cat experiment once more. When the charge is in one of the (say) two locations in superposition, then it generates a disturbance in the electromagnetic field that propagates outward from the charge's location. The same happens when the charge is in the other location, only the disturbance originates in that location instead. Because the charge is in a superposition, the two field disturbances also exist at the same time (just like the cat).

The situation has the same logic as that of observers making quantum measurements: here, the field "measures" where the location of the charge is by entangling itself with the charge's position. And if charges affect the field by being in multiple places at the same time, we can't possibly treat the disturbances classically. In other words, we'd need to describe the harmonic oscillators that make up the field with q-numbers. But—and this is the problematic *but*—making this change from c-numbers to q-numbers in these modes of the field is thought to be a gauge transformation. In other words, the conventional way we do quantum field theory of the electro-magnetic field says that we won't be able to tell the difference between the two ways of doing things when describing the static attraction and repulsion between charges.

The basic idea of gauge invariance is that our choice of coordinates must not affect our conclusions about what's really going on. For instance, it would be funny if going to a bank with dollars, changing them into pounds, and then back into dollars enabled you to buy a great deal more or less with your dollars—no matter which currency you use, you should have the same amount of money. Likewise, it would be strange if merely relabeling the locations of our detectors actually caused a photon to be detected by a different detector! The facts of the universe must remain the same, no matter what language we use to express them.

The change from c- to q-numbers describes the so-called ghost modes of the electromagnetic field. The word *ghost* suggests that the apparent differences between those two descriptions are an artifact of using a specific gauge and cannot change the fundamental physics. Contrary to mainstream thought, I believe that they do and that this change can be detected.

I'd like to describe an experiment, our fifth portal into the future physics, that could actually tell the difference and show that making the change from c- to q-numbers is not a (classical) gauge transformation. Also, I'd like to tell you why no one has thought of this idea before. It has to do with one of the information principles we've discussed before—quantum local tomography.

Local tomography tells us that in order to confirm that a system is in a superposition of states (say, in two different physical locations) we do not need to bring these two states into one and the same location. We could, and that will do the job most of the time, but here we'd like to confirm the superposition by making measurements in one location and, separately, in the other. The principle of local tomography says that this can be done provided that we can repeat our measurements over and over again—which is how experiments are normally done anyway.

Ghosts can be detected because when a charge is in a superposition of two different places, it becomes entangled with the ghost modes of the electromagnetic field. As we said, this is the explanation of the static repulsion between like charges (such as two electrons). The entanglement between the charge and the ghost mode vanishes if we bring the charge back to the same place in order to perform quantum interference. This is a bit like converting the dollars into pounds and back into dollars, as you simply get back to the same place from which you've started.

The key is not to do an experiment in which the two positions of the electron are reunited, but instead to measure the state of the charge by independently probing each of the locations in which the charge exists. No present treatment of quantum field theory takes this possibility into account, which is why the experiment I am describing has been missed by virtually everyone working in physics. But the results of such an experiment could very well lead us through a portal into brand-new physics.

When we perform local tomography on that superposed charge, then, if the entanglement with the ghost modes is real, the extent of the electron superposition will be diminished. If we only have a single electron, the entanglement is small but still detectable. With a larger charge in a superposition, it becomes larger, and quadratically so (two electrons lead to four times the amount of entanglement, three electrons to nine times the entanglement, and so on). Ultimately, if the size of the charge is large enough, the entanglement with the ghost mode would be maximal, just like the Schrödinger's cat experiment. In that case, the superposition of the electron would completely vanish. The electron and the ghost mode would be exactly like the atom that is and isn't decayed and the cat that is both dead and alive.

However, just with other key experiments discussed in this book, we could also fail to detect entanglement in this experiment. This would suggest that static interactions are truly classical after all. Such a result would be a return to some kind of Newtonian action at a distance, although it would not lead to speeds faster than light. Still, if quantum physics continued to be local, it seems to me that the ghost modes must be quantum.

We could ask the same question and conduct the same experiment for the static gravitational forces as well. My bet—as you might have gathered—is that entanglement would be detected, forcing us to acknowledge that the ghost modes of gravitational fields are real and quantum, too. However, when it comes to ghosts in gravity, maybe there's a bigger fish to fry.

IS SPACETIME A GHOST?

Someone witty once said, "It's hard finding a black cat in a dark room," and then added sarcastically, "Especially when there is no black cat." This metaphor of a nonexistent black cat in a dark room is frequently used by atheists to argue against religion (God being the cat), but here I want to argue against the view that space and time need to be quantized. And, yes, you've guessed it—I will argue this by claiming that there is no such thing as spacetime, that spacetime is a glorious nonentity.

Remember that quantizing something means treating its basic components as q-numbers and not just as ordinary c-numbers. In other words, if space and time became q-numbers, this would imply that we cannot simultaneously measure, say, different spatial components and/or time. It would be the same as the uncertainty that exists between the position and velocity of a particle. Heisenberg says you cannot know/specify them at the same time. Could the same be true of space and time? The more accurate your clock, the less well you can tell where it is. It sounds funny, but in physics we are by now well used to many counterintuitive concepts.

Still, I'd like to claim that the universe is not such that space and time are q-numbers, for the simple reason that there are no space and time. They are our convenient labels, conventions, and simplifications, but they don't really exist. There are no elements of reality that correspond to space and time!

But what about Einstein's theory of relativity? Isn't that all about bending space and time? Well, that's what I'd like to tell you about here, and it's a view of relativity that has nothing to do with spacetime!

Before you lock me up in an asylum, let me justify my statement a bit. A while back I argued that gravity, as in Einstein's general relativity, will prove to be quantum (the BMV experiment will settle that question), but this will not mean that spacetime is quantum. It will mean that the *gravitational field*, not *space and time*, should be described by q-numbers.

But now I would like to talk about special relativity (to remind you, this is the physics of space and time in the absence of gravity). We all remember the experiments showing that moving clocks tick more slowly than stationary ones—the phenomenon of time dilation. This was Einstein's prediction, and it has been tested extensively. So there is no doubt that time dilation is real.

The main question is why time dilates. The standard view is that motion bends space and time. In other words, by moving we slow down time. Here I want to argue against this view that moving clocks, or anything else moving, can even affect either time or space. But if time and space are our convenient myths, then what is it that gets dilated? The answer is electromagnetism.

Matter is held together by the electromagnetic forces. Atoms exchange photons, and this process of exchange attracts them into larger molecules. Molecules, in turn, exchange photons, a process that binds them into even bigger molecules. And so on. All clocks and all rulers that are used to measure time and space are held together by electromagnetism (I am ignoring the forces inside the nucleus, which are not relevant, as well as gravity, which is too weak at this level).

The punch line now is the fact that different components of the electromagnetic field transform in the same way as space and time do. The four components of the so-called electromagnetic vector potential, the key quantity characterizing the electromagnetic field, transform when in a moving frame, exactly as space and time do (the component in the direction of motion contracts, while the component that does not have direction dilates). That leads us to think that time dilates, whereas it is the electromagnetic field that changes, leading to an effect as though the time itself has changed.

If it's true that spacetime is not fundamental, then approaches to quantum gravity such as loop quantum gravity have to be tossed out. If the BMV experiment succeeds, it would support the idea that it's the components of the gravitational field that are q-numbers, not space and time (although the result would also be consistent with quantized spacetime).

Two big questions follow naturally. Can we experimentally discriminate the views according to which either spacetime or electromagnetism is responsible for time dilation (and another consequence of relativity, length contraction)? Second, why are the laws of electromagnetism such that they lead to these particular transformations, the so-called Lorentz transformations, of its different components? At present, we have to admit that we don't know. According to Maxwell's equations, which are the most fundamental description we have of the electromagnetic field, this is just so. Maxwell's equations also say that the speed of light is the same for everyone, and this too is a mystery. So while I've argued that the mystery of space and time might be gone (i.e., they might not exist at the most fundamental level), other mysteries have immediately popped up. But such is the spirit of the scientific inquiry: you solve one problem only to create two more.

What does this imply about general relativity? As I've discussed, the general relativistic principle is that no matter what coordinates we use to express the laws of physics, we should obtain the same result. General invariance says that having two different coordinate frames ought not to change the underlying physics. This is important because it tells us how gravity must interact with matter, and it revolves around the same gauge principle as electromagnetism.

Imagine that you are holding a stick at both ends, and that you want to freely move the ends any way you please. Clearly the stick would not be able to remain rigid, but would have to change shape as the ends are twisted and turned. The only way that would not happen is if there was a force between the atoms of the stick holding things together. Of course, there is such a force, and it is the electromagnetic one. Molecules of the stick are coupled together through the electromagnetic field. In short, insisting that the stick stays invariant (i.e., the same stick) when we wiggle its ends "forces" us to introduce the electromagnetic field.

It is the same for gravity. If we allow arbitrary special relativistic change of coordinates throughout spacetime, then there is only one way—which includes a couple of simple and natural assumptions that need not concern us here—to preserve the way by which matter couples to gravity. Gravity holds it all together when each of the points waves independently and arbitrarily.* And in none of this do space and time need to be fundamental. They might just be our agreed-upon coordinate labels on which all quantum waves exist. This could be a different way in which ghosts enter physics: not as modes and particles that are not directly detectable, but as labels that are convenient to use but don't have any fundamental significance.

BYE-BYE, VIRTUAL PARTICLES

In almost all standard textbooks, static interactions are described using the language of "virtual particles." They say that an electron can emit a photon with negative energy,

---

* In fact, all four fundamental forces are currently understood as arising due to some invariance or other.

meaning that the photon borrowed the energy from the electromagnetic field. This clearly violates energy conservation—that's why the photon in question is called virtual. However, so this account goes, energy conservation can be violated so long as the energy is returned quickly enough. "Quickly enough" means that the time ought to be smaller than the Planck constant divided by that energy. Then the virtual photon is absorbed by another electron, which is how the energy is restored, and this gives rise to the repulsion between the electrons. This kind of language exists for other phenomena as well, such as quantum tunneling or even in chemistry when we describe how atoms bond into molecules. Virtual particles are very closely related to ghosts—one can think of them as excitations of the ghost modes.

I think that this account of what happens is wrong, and that hypothesizing about virtual particles is a desperate attempt to insist on a classical description for a process that is ultimately completely quantum mechanical. It's high time we exorcised them.

As we have been emphasizing, waves are the fundamental entities in quantum mechanics, not particles. Given this fact, if we should not appeal to virtual particles, what is the appropriate way to describe interactions in quantum physics?

When two systems—atoms, electrons, waves on different sides of a wall, and so on—are not interacting, each of them has its own stable energy levels, such that the systems will stay in the same state for a long time. However, once the systems are able to interact (for instance, if two atoms get close enough to each other so that their electrons can hop from one to the other), then the new stable energy levels are different from the original ones and are actually superpositions of the original (non-interacting) energy states.*

Put simply, as systems interact, they can find and remain at stable energy levels that they never could have in isolation. This is how we get dynamics, and how a wave that is confined to one side of a wall can dynamically end up on the other side of the wall. This is called quantum tunneling and is sometimes mysteriously presented by saying that a particle that doesn't have enough energy to go through a wall ends up doing so. This is misleading, since it suggests energy conservation violation, which is certainly not what happens in quantum mechanics.

Sometimes, too, we read that the tunneling particle has a negative amount of energy while inside the wall. This is the same as the language of virtual photons and is simply not the appropriate way of understanding what is going on. It is because we insist on waves being particles with well-defined energies, positions, and velocities that we get into trouble. The cure is to acknowledge that everything ultimately is a q-wave; when we do this, the account of any of these phenomena is then paradox-free. This wave nature allows particles to have a spread-out extent that goes beyond the size of the wall and therefore permits tunneling.

The takeaway is that when we stick to using *only* q-numbers in our descriptions and calculations, we never encounter any inconsistencies. Apparent paradoxes arise only when we invoke the half-quantum and half-classical description. When

---

* In fact, some of these new energies can be lower than the original ones, which is basically why the atoms are—quantum mechanically—drawn to one another. They tend to go to the lower energy state.

we replace all c-numbers with q-numbers, the world makes much more sense!

––––––––––

We've now seen that much of the language we use to communicate quantum phenomena is in fact based on the remnants of classical physics. These remnants are helpful to the degree that they aid us in some calculations, but we should be cautious not to take such classical language and machinery too seriously, since doing so might prevent us from making real progress.

I keep hearing that the twentieth century was the century of physics, because of quantum physics, relativity, the atomic bomb, and all of the other fundamental discoveries that have radically changed the world we live in. I am also told that—sadly, for a physicist such as myself—this is the century of biology (look at all the advances in molecular biology, genetics, and so on). Well, I hope that my book has gone the distance in convincing you that the twenty-first century, too, will be the century of physics.

Having presented my evidence, ladies and gentlemen of the jury, now I'd like to stick my neck out and tell you what I think that next theory might entail.

TO INFINITY AND BEYOND: QUANTUM PHYSICS, ALL THE WAY DOWN

So, what now? I have outlined five portals that can lead us to new physics beyond quantum theory and relativity. I have led you to the edge of a cliff and we are now staring into the unknown. This is where we need to become creative and summon all our imaginative faculties. The content of the theory that comes next will reflect the outcomes of the experiments I have described in these pages. However, regardless of its specific content, we will need to stretch our creative minds in order to imagine what kind of theory it will be. What mathematical tools will it harness? Will it be another dynamical law, like the

Schrödinger equation, or will it take a different form—maybe it will be composed of principles, like those underpinning thermodynamics and relativity? Perhaps our analogies, our present-day language, and even our current, limited modes of perception are a trap preventing us from grasping properly the content of the new theory. Like a general on a complex battlefield, we need to keep all these possibilities in mind and prepare accordingly.

HOW MUCH OF MATHEMATICS IS REALLY JUST PHYSICS?

I'd like to tell you about my late friend Professor Peter Landsberg, who was a thermodynamicist (this is how he liked to call himself to emphasize his area of specialty within physics). I used to meet him at the London Athenaeum in the good old days when cigar smoking was still allowed indoors (yeah, I agree that it's not good for one's health, but I still miss it).

We would sit in one of the rooms of the club, order a drink, and—while lighting our cigars—start a discussion about some physics topic or other. Among the quirky rules of the Athenaeum was the fact that you could not enter the club without a jacket and a tie, but if you did show up underdressed, the club would fit you with the appropriate attire (as I experienced myself the first time I went there; it's straight from the "Suit you, sir!" routine of the sitcom *The Fast Show*). Funnily enough, the Athenaeum also had some old-fashioned rules according to which you were not meant to talk shop at the club. If we began to write our calculations down on a piece of paper, there was always someone there to tell us off for violating the rules of the club.

Peter was fond of my field, and I have likewise always been mesmerized by thermodynamics. My PhD thesis was based on using methods similar to the ones in thermodynamics to quantify quantum entanglement. It turns out that entanglement can be quantified with the same entropy that is used in thermodynamics and information theory.

I also enjoyed Peter's stories about some of his exchanges with other famous physicists, such as the time he proved Pauli wrong on ergodicity (there is a letter from Pauli to his collaborator Markus Fierz basically admitting that Landsberg's criticism of their work is fully justified—you can find it on the web if you search on "Pauli, Landsberg, H-theorem"). Proving Pauli wrong was not easy, and the story reinforced for my much younger self the important lesson "Take no one's word for it."

Peter was very passionate about thermodynamics (I laughed when I learned that his son's first word was "Carathéodory," the last name of the mathematician whose formulation of thermodynamics Peter helped rescue from obscurity), and our discussions were always exciting. We even wrote a paper together on information theory and thermodynamics (introducing an entity now called the Landsberg-Vedral entropy). One of our discussions in particular will always stay in my memory. During our meeting, Peter told me something absolutely mind-blowing: the fact that the arithmetic mean is always bigger than or equal to the geometric mean is a direct consequence of the second law of thermodynamics.

I initially felt that this cannot possibly be right. What on earth is the connection between the two? The second law says that the disorder in the universe can never decrease and usually goes up until a maximum is reached. The arithmetic and geometric means are just statements about statistics, different ways of talking about averages. *Yeah, of course Peter would say something like that*, I thought, *as he is an ardent thermodynamicist and probably thinks everything is a consequence of the second law of thermodynamics.*

But then he proceeded to prove it to me. I will illustrate it with two numbers, though the generalization is immediate. Take two non-negative numbers, a and b. Their arithmetic mean is $(a + b) / 2$, while their geometric mean is $\sqrt{(a \times b)}$. Now, why does the second law say that the former is greater than or equal to the latter?

Here is where Peter's beautiful argument comes in. Take two identical objects, but at different temperatures—say, temperatures a and b. What final temperature will they reach if we put them together to allow for heat exchange between the two? If we just put them in contact without doing anything else, the final temperature will simply be the arithmetic mean, $(a + b) / 2$. This can be understood by way of the first law of thermodynamics, which stipulates that the initial and the final energies have to be the same.*

However, if we conduct the same process reversibly,** then total entropy is conserved. The relationship between entropy and temperature is described by a well-known thermodynamic formula. The intuition behind the formula is that the higher the temperature of the system, the more its constituents "jitter." This results in a state of higher disorder—in other words, greater entropy. This tells us that $\log a$ plus $\log b$ is equal to $2 \times \log f$, where f is the final temperature in this case (and $\log$ is the natural logarithm). Rearranging, we obtain $f = \sqrt{(a \times b)}$.

---

* Energy is just the Boltzmann constant times temperature.

** This means slowly enough that no extra heat is being dissipated throughout.

Now comes the punch line. The final entropy in the first process must be higher than in the second process. This is because the second process is reversible, and the second law says that entropy does not increase for reversible processes. So the entropy/disorder of the first process must be greater (as the first process is dissipative and therefore increases total disorder); that is, $(a + b) / 2$ is greater than $\sqrt{(a \times b)}$. They are equal only if $a = b$, in which case the temperatures are already equal to start with and nothing changes when we put the systems in contact.

I was speechless. Just think about it. If we lived in a universe in which the arithmetic mean was smaller than the geometric one, then this universe would also violate the second law of thermodynamics! Our lives would be like Benjamin Button's, going in reverse from old age to infancy.

These kinds of thoughts prompted me to look for other instances in which the laws of physics implied mathematical relations. They made me also begin to form the idea that maybe the whole of mathematics just follows from the laws of physics! (I am sure all my colleagues in the math department would strongly disagree.)

This view, that physics constrains mathematics, goes against the view of Max Tegmark, who argued for the opposite in his book *Our Mathematical Universe*. There, Tegmark suggested that mathematics exists independently of physics and that different kinds of mathematics are all instantiated in different universes. We live in one of those universes, where one kind of mathematics governs physical processes, but in other worlds other kinds of physics would be happening that correspond to different mathematical structures.

What I am suggesting, on the other hand, is that the laws of physics come first and constrain the kind of mathematics we get. If you keep asking why $1 + 1 = 2$, you will ultimately get to the laws of quantum mechanics. Integers exist not because God created them (as Leopold Kronecker famously said) but because the fundamental laws of physics are those of quantum mechanics.

Of course, we know that many things we observe around us are the way they are because physics is the way it is. A random example comes to mind: spiders would not be able to climb up vertical walls if it weren't for a quantum-generated force between their hairy legs and the atoms constituting the walls. This force is so strong that it effortlessly counters the gravitational pull downward (yes, spiders, too, owe something to quantum physics). Almost anything we see around us, not just the phenomena in the micro world, is a consequence of the simple laws of quantum physics and gravity.

However, mathematics is sometimes thought of as having a life of its own. Its existence somehow seems independent and as real as what we see around us ($2 + 2$ is surely always 4 independently of our world, right?). As I alluded to earlier, Plato thought that mathematics is the true reality, and that the real world is only an imperfect copy of the immutable mathematical entities of the (what we now call) Platonic world.

Well, my friend Peter Landsberg had the courage not just to challenge this view but even to turn it upside down. His creative way of viewing (at least some parts of) mathematics as following from the laws of thermodynamics was certainly an inspiration for me. One cannot help but wonder: Do we have integers ultimately only because everything is made up of quanta? Is geometry the way it is because of Einstein's general relativity? Computers can exist in both the classical and quantum worlds, but it is only in the quantum world that they can do certain computations much faster. Is anything in mathematics, even some totally esoteric result that no one has seen any use for in the real world, ultimately a consequence of the laws of physics?

This view is definitely pleasing to a physicist. But it would also help us address the incredulity Wigner expressed at what he saw as "the unreasonable effectiveness of mathematics in physics." Mathematical effectiveness would not be unreasonable at all, since ultimately every mathematical truth would depend on physical truths.

I wanted to talk about this because it is also possible that a new kind of mathematics is needed in order to discover the next theory of physics. This would be in the same spirit as Newton inventing calculus in order to express his laws of motion.

Of course, we mustn't ignore the relationship between philosophy and physics, either. Philosophy, physics, and mathematics are all inseparably connected with one another. And they are all underpinned by our current modes of expression and by the language we use to describe the universe. Language reflects our perception of physical reality, which is primarily shaped by the macroscopic phenomena around us. It has also evolved to incorporate some of the complex ideas that emerged with each scientific revolution; however, it still remains limited in its capacity to express the vision that contemporary physics provides us with. This is why parallels with familiar objects and situations come in handy, but they can also be deceptive.

ANALOGIES

When we communicate deep and counterintuitive ideas in physics, we often use everyday analogies in order to make them easier to understand. Nowhere is this more the case than in quantum physics. On top of that, quantum particles, such as atoms and photons, can (unfortunately for us) behave in such bizarre ways that there are simply no counterparts in the macroscopic classical world that we can rely on as accurate and relatable metaphors. So I think it is important to ask, philosophically, what we risk and what we might miss when we work with analogies.

Within quantum mechanics, the quantum concepts we find difficult to understand (such as the superposition principle, quantum measurement, the existence of entangled quantum states, and the Heisenberg uncertainty relations) are all related to one another and are simply a consequence of the fact that we use vectors to describe quantum states and matrices to describe what we can observe.* Of course, putting things in this mathematical way does not really illuminate what's going on physically when we do quantum experiments. Even to a physicist, mathematics is simply not enough— we'd like to actually understand quantum reality. Hence our analogies.

Let's start with superpositions. Quantum superpositions are sometimes analogized to thinking. Thinking is frequently fuzzy, in the sense that our thoughts are not always fully formed. Sometimes we even say that we are of two minds about something. Of course, this really only means that we are aware of two or more options and are going back and forth between them. Our thoughts don't really exist in a superposition but, I would guess, are stored in different parts of our brain at the same time. It is apparently not the case that one and the same

---

* Quantum states can ultimately be expressed as matrices, too.

neuron is quantum superposed (although this has not been ruled out, either).

Quantum measurement is counterintuitive because it supposedly "collapses" the state of the system being measured from a superposition of states to one definitive state. The unusual thing here is that a quantum measurement changes the state of the system. Classically, when we measure the speed of a car, we do not expect our measurement to change the speed we are measuring (otherwise we would have a great defense against speeding tickets). But quantum measurements are like strict army officers who do not allow any room for uncertainty (this analogy is courtesy of physicist Bill Wootters).

Lieutenant: "Did I make myself clear, soldier?"

Soldier (very nervous): "I—I—I am not so su—sure. You did mumble a bit, and I may not have underst—"

Lieutenant (getting angry): "*I asked you something, soldier: did I make myself clear?*"

Soldier: "*Sir, yes, sir!*"

Lieutenant: "*Have I ever been unclear, soldier?*"

Soldier: "No, sir, never. You've always been *crystal* clear, *sir!*"

Here again, the analogy is between an initially indecisive soldier and a quantum superposition. Of course, being unclear is not the same as being superposed (for superpositions are very clear and definitive quantum states in the sense that we know what all of the branches of the superposition look like, even if we can't predict which branch we'll end up in), but after the measurement event, the quantum system does indeed "forget" its previous state (in the same way that the soldier is forced to admit that the lieutenant has never been unclear).

Arthur Eddington, the famous astronomer who measured the bending of light and thereby confirmed Einstein's general theory of relativity in 1919, compared quantum measurements to footprints in the sand. They erase anything else that was there before, but they also don't last long and can be erased by another event such as a rising tide (in analogy with the subsequent quantum evolution of the system after it has been measured). The sand analogy seems especially apt due to the impermanence of the quantum world, where detectable particles can pop in and out of existence and can be converted into other particles at mind-boggling speeds.

But measurements are not forgetting or collapsing; measurements in quantum physics are really just entanglements between different systems. And quantum entanglement has had more analogies associated with it than anything else. My colleague Charles Bennett, one of the pioneers of quantum information, is fond of the following one. Entangled quantum systems are like hippies, says Bennett. None of them have any strongly held beliefs, yet they all feel in perfect harmony with one another. Ha, ha, ha. I don't know if hippies happen to be like this (Bennett does—he is from the baby boom generation), but entangled states certainly are. When two particles are quantum entangled, neither has a well-defined state, but when one of them is measured to be in one state, the other one "jumps" to its corresponding state.

Sometimes we talk about identical twins to explain entanglement. The twins could be separated at birth, but as soon as we meet one of them and notice that their eye color is blue, we know that the twin must have the same eye color (no matter where they are). However, the limitation of the analogy here is the fact that, unlike the twins, quantum particles could be in superpositions of two or more different colors. When two photons are entangled in their frequency, then we can measure one of them to be in a superposition of blue and brown, upon which the other photon will be in the same superposition of blue and brown. Clearly this has no analogue with twins, whose eye color cannot be in a superposition.

Quantum entanglement is a form of correlations that has no classical analogue. Schrödinger called entanglement "*the* characteristic trait of quantum physics," and of course he introduced the famous cat to illustrate it. Schrödinger's cat, however, is not an analogy for, but a consequence of, quantum physics if applied to the macroscopic domain. Sometimes we hear it said that we don't know if the cat is dead or alive until we open the box. It is true that we don't know, but not because it is both dead and alive. It is because the cat is entangled with the atomic decay and the poison. Both branches of the superposition exist, although, as we saw, Bob will only ever experience one of the outcomes in each branch.

Some people are bothered by the fact that particles can be entangled across vast distances. To others, however, this suggests that there is a certain underlying unity of everything in quantum physics. Entangled particles should not be thought of as separate entities, even when spatially distant. David Bohm, one of the quantum pioneers, talked about "wholeness and the implicate order." He gave the following analogy to illustrate this interconnectedness. Imagine that there is a fish tank with a single fish inside, of a kind unlike any fish you've ever seen. Furthermore, imagine that there are two cameras pointing at the fish, one head-on and the other sideways. All you see are two well-separated screens on which these two images are projected. Now, when the fish moves, you will see the images change in a way that seems magically coordinated, since you think that each image corresponds to a distinct fish. In fact, you might even start to suspect that this phenomenon violates the theory of relativity, as the motion between the "two" fish seems to coordinate instantaneously. Bohm thought that entanglement points to an even deeper hidden reality behind the entangled systems. Here Bohm follows in Plato's footsteps and his famous cave analogy: the power of the analogy is not just to give us a feeling for something but to instruct us that perhaps we should explore what lies beyond.

The uncertainty principle tells us that certain properties, such as the position of a particle and its velocity, cannot be measured to arbitrarily high accuracy by the same measuring device. The better we measure the position of an electron, the less we can be certain about its velocity. Here we are back to the process of thinking as a good analogy. This quote comes straight out of Bohm's classic textbook on quantum theory: quantum uncertainty "is analogous to asking for a detailed description of what we are thinking about while we are reflecting on some definite subject. As soon as we begin to give this detailed description, we are no longer thinking about the subject in question, but are instead thinking about giving a detailed description. In a similar way, when an electron is moving with a definable trajectory, it simply can no longer be an electron that has a definite energy."

Bennett similarly talks about quantum states being like dreams and telling someone about our dreams being like a quantum measurement. Our recollection "collapses" the dream, and after the recollection all we remember is our own story and not the dream as it was. In this sense, dreaming and describing a dream are two things that cannot happen at the same time and are subject to an uncertainty principle. Of course, we do not understand processes in the brain well enough to really explain how thinking and dreaming work—in contemporary neuroscience, there are no laws like there are in physics. One day, when we understand consciousness as well as we understand physics, perhaps our analogies will improve, too.

For every quantum analogy, there is also one in general relativity. While gravity is the curvature of spacetime, it is hard for us to imagine four dimensions, so we instead think of a three-dimensional sphere. We've talked about how curved paths on a two-dimensional map look like straight lines on a sphere (and vice versa). Then there is also an unusual possibility that spacetime could be curved without the presence of matter. In his *ABC of Relativity*, Bertrand Russell presented a metaphor in which there is a tiger in a town square and everyone is running away from it in panic shouting, "Tiger, tiger!" Imagine you were a passerby who saw people running away in all directions from the town square but could not see the square itself. You would think that there indeed was a tiger there. Russell's point is that there need not be even though everything would look as though there was. The same, he thought, was true of gravity.

The fact that different observers perceive space and time differently—remember space contraction and time dilation— has been subject to many analogies. For instance, how is it possible that when we are moving at different speeds my time slows down with respect to yours but, due to the inherent symmetry of the situation, your time also slows down with respect to mine? Likewise, how can your distance contract with respect to mine and vice versa simultaneously? Both seem like contradictions. However, Bohm had a simple way to resolve the problem by way of analogy: when two people are approaching each other from a distance, each looks smaller to the other one. No problem, no contradictions.

All this shows the complexity of rendering with words what the mathematics behind physical laws captures in very few symbols. One cannot exist without the other, and our language too has to evolve together with our improved understanding of the laws of physics. You are now prepared for this joke:

> Heisenberg and Schrödinger are in a car, and a
>   police officer stops them for speeding.
> The police officer asks, "Do you know how fast you
>   were going?"
> Heisenberg replies, "No, but I know exactly where
>   I am."
> The police officer tells him, "You were doing 55 mph
>   in a 30 mph zone."
> Heisenberg throws up his hands and shouts, "Great!
>   Now I'm completely lost!"
> The police officer thinks this is suspicious and orders
>   him to pop open the trunk. The officer checks
>   it out and calls over, "Do you know you have a
>   dead cat back here?"
> "We do now, you dimwit!" shouts Schrödinger.

PRINCIPLES: *VIA NEGATIVA*

When thinking of the future theories of physics, one can contemplate something far more radical than adopting new mathematics or a novel kind of language. One can even think of changing the way fundamental laws are expressed. Here is where principles may come in. John Wheeler emphasized that there could always be some simpler principles that explain the existing laws of physics as we know them. For instance, in relativity we measure distances between events by summing up the squares of the spatial distances (as we would do in the Pythagorean theorem), but then we subtract the square of the time difference before taking the square root of the whole thing. In this formulation, time seems to be an imaginary (squaring an imaginary number yields a negative real number). Why does relativity rest on this kind of weird distance measure?

Your average physicist would probably just say, "This is the way it is. That's what relativity teaches us, and here is all the evidence in favor of relativity." However, Wheeler was different. He always looked for a different kind of explanation, something that involved a deeper physical principle, one that was easier to understand and that implied other things we knew.

If we solve the simplest equation for the dynamics of the universe that is based on the metric in which time is imaginary (it is the metric that tells us how to calculate distances in spacetime), we obtain a solution that has a clear beginning and reaches a maximum size. If, on the other hand, we use the metric in which time is additive and real in the same way that space is, we get dynamics in which the universe has a smallest size but no beginning at all. Therefore, if we postulate a principle stating "The universe must have had a beginning," this automatically leads us to the metric characteristic of special relativity (and which also holds locally at every point in general relativity)—namely, the imaginary aspect of time.

Now, I agree that the principle "The universe had a beginning" may not be superior to the principle "The universe had no beginning." But my point was just to illustrate the idea of looking for some simple underlying concepts that could ultimately fix all of physics as we know it.

What principle might underlie the next revolution in physics?

I am much more sympathetic to the Eastern religions than to the Abrahamic ones. However, there is one branch of Christianity that resonates with me more than the rest of it put together (admittedly, this might not be saying much in my case). It's called the Negative Way, or the Via Negativa.

This version of Christianity was practiced by the Cappadocian Fathers, fourth-century Christian monks who lived in a stunning region of Turkey north of Istanbul. They resided in caves and practiced a rather ascetic and spiritual form of existence (you can now stay in one of the caves as a tourist, and I believe Angelina Jolie did just that in her role as Lara Croft in *Tomb Raider*). They developed what is frequently called a mystical form of Christianity, though I think that this is an inappropriate label. Rather, I find their approach to be one of the most rational of religions (let me put it even more strongly: if you forced me to be a Christian, this denomination would be what I'd go for).

The reason I think they were rational is that their philosophy, the way they thought about God in general, in many ways resembles the way we scientists tend to phrase things about Nature.

Scientists cannot tell you what things really are like, but we can quite confidently say what they are *not* like. The Earth is *not* flat. The Sun is *not* at the center of the universe. You *cannot* measure the position and speed of an electron at the same time. You *cannot* use all the energy to do useful work. You *cannot* decrease the entropy of the universe. You *cannot* travel faster than the speed of light. You can *never* do any experiment to detect uniform motion. Science doesn't yet know what is ultimately possible (and it may never know), but we are quite sure about what cannot be done. Science is the Via Negativa par excellence.

As we saw earlier, Einstein once said that the more prohibitive the law, the better it is from a scientific perspective. This is because more prohibitive laws rule out larger numbers of (wrong) possibilities. And when you've ruled out as much as you possibly can, whatever remains, however weird it might be, must ultimately give you the final truth (I am sure there is a Sherlock Holmes quote hiding in here somewhere).

The Cappadocian Fathers would say similar things about God. You cannot say that God exists. Equally, you cannot say that God doesn't exist (both statements sit really well with atheists). You cannot say that God is infinite (or indeed finite) or that God created the universe. And so on and so on. Now, it would be a stretch to say (and I am *not* saying it) that the Cappadocian Fathers would have arrived at quantum physics and general relativity this way, but they certainly did influence many things in medieval philosophy that may have ultimately led us to Kepler, Galileo, Newton, and the rest. Who knows?

I don't think that the Via Negativa will lead us to a new theory in physics. But it is still appropriate to emphasize that you should have the same negative attitude toward most of the things you read about physics in the popular press these days (some of you would probably add that this attitude should apply to everything in the press, not just physics).

More serious, but still negative, is constructor theory, due to Deutsch and developed by Marletto—our modern-day Basil the Great and Gregory of Nyssa—which states that all the laws of physics should ultimately be phrased in terms of principles that tell us what tasks can and cannot be performed in our universe. It is all about which physical transformations can possibly be brought about by an arbitrary constructor, and which physical transformations cannot possibly be brought about by an arbitrary constructor. The theory splits all of physics into possible and impossible transformations. The idea is to have a finite set of principles that tells us what cannot possibly be constructed. Then whatever remains is indeed our physical reality. When I think about this very appealing idea, Carathéodory's formulation of the second law of thermodynamics always comes to mind.

Carathéodory said that in the vicinity of any adiabatically possible process (a process that can occur without the system in question exchanging heat with the environment), there exist processes that are not adiabatically possible. Here, "in the vicinity" means that the processes are close to one another in the sense that a small change in one of them will lead to the other process. The Manchester brewer (and physicist) James Joule knew this well. It is easy to heat up beer in a barrel by stirring it, but no adiabatic action exists that could do the reverse (a shame, otherwise I wouldn't have to pay for the fridge to keep my beer cool).

Extrapolating from Carathéodory, and channeling Deutsch and Marletto, we might hazard a guess at the most prohibitive constructive theoretic law of physics: in the vicinity of any physically possible task, there are tasks that are impossible even by the standards of the universal constructors (constructors that can cause any possible transformation).

Sounds prohibitive, sure, but is it prohibitive enough? More importantly, can we even derive the existing physics from it? Even more importantly, will it give us any *new* physics?

THE PRINCIPLE OF
"NO MINOR TWEAKING"

The claim that quantum physics is somehow "tighter" than classical physics has been advocated a number of times by many different people. A tighter description of reality usually is considered a better explanation for the observable phenomena, but what does "tighter" actually mean? Roughly speaking, here it means that the mathematical structure underpinning the laws of quantum physics is such that a small change in one of its laws has a dramatic effect, to the point of making the physics described by the slightly modified theory completely nonsensical.

For instance, the quantum laws of dynamics are linear. This means that if a particle evolves in time by moving from place A to place C and from place B to place D, then if the particle is in the superposition of being in place A and place B, it will evolve into a superposition of being in places C and D.

The linearity of quantum physics leads to the no-cloning theorem,* quantum entanglement, Schrödinger's cat, and all the important recent applications such as quantum cryptography

---

* Loosely speaking, the no-cloning theorem says that it is impossible to make an identical copy of an unknown quantum state.

and quantum computation. But it has been questioned by various people, mainly due to their belief that quantum superpositions might collapse at some point. People have proposed nonlinear modifications to quantum physics, but those modifications usually suffer from grave difficulties. Nonlinearity means that two different states could deterministically evolve into one and the same state. If quantum mechanics was simply linear, then when we measure a particle initially in a superposition of places A and B, we *might* observe it at B. But if quantum mechanics was nonlinear, then when the particle is in a superposition of places B and C and we measure it, we *always* obtain B. This is possible according to some of the proposed nonlinear tweaks of the Schrödinger equation. It simply means that what happens to the states individually is very different from what happens when they are in a superposition.

Making the Schrödinger equation nonlinear, however, gets us into trouble with some other principles we hold dear. For starters, it could allow us to communicate instantaneously at arbitrarily large distances. It also could lead to a reduction of the entropy of a closed system, thereby violating the second law of thermodynamics.* In other words, even a tiny nonlinear modification to quantum dynamics immediately gets us into trouble with both relativity and thermodynamics!

In contrast to the laws of quantum physics, the laws of classical physics could be modified in a number of different ways without obtaining any disastrous inconsistencies. One could, for instance, say that classical physics is full of nonlinearities similar to the one I described, but this does not necessarily

---

* This is because, all else equal, having many possible micro states for a given macro state is more entropic than fewer possible micro states for a given macro state, and nonlinear dynamics could lead to the reduction in states.

lead to any huge violations of other principles of physics. In that sense, quantum physics is tighter than classical physics.

However, I am not really sure how exactly to compare classical and quantum physics when it comes to tweaking their basic principles. If we were to compare them in this way, we would first have to decide on what corresponds to the same amount of tweaking in each theory. Second, we would have to measure the impact that the same amount of tweaking has in each theory in order to compare them. We don't really know how to do either of these things.

The way things happened historically, both quantum physics and Einstein's relativity were indeed huge departures from Newtonian physics. We definitely didn't get from Newton to Einstein or to quantum physics by introducing a small modification to Newton's laws. The leaps were ultimately so big that what we considered the fundamental entities in physics underwent a complete revision.

In classical physics, for instance, force is a fundamental entity. In quantum physics, force is replaced by energy. In Newtonian physics, gravity is a force. With Einstein, it becomes a geometry of space and time—completely different from the original Newtonian conception.

It is perfectly clear that new theories in physics are huge departures from the old ones, although they always contain the old theories in some special limits. But are quantum physics and relativity the tightest that the laws of Nature could be? If not, could "tightness of the laws" be a principle to guide us to the ultimate laws of Nature that lie beyond quantum physics and relativity?

We could phrase this principle by saying that "the ultimate laws of Nature are those that allow no tweaking whatsoever."

In other words, if we have an allowed process in the universe (such as the transformation of water into ice by cooling), then the smallest change to this process would lead to an impossible process—that is, something that can never occur in the universe (for instance, that by further cooling ice we get back water).

This way of talking about the laws of physics echoes Einstein's comment that the best laws of physics are the most prohibitive ones. Take his first principle of relativity, which he borrowed from Galileo and Newton: no experiment whatsoever can determine whether or not we are moving at a constant speed. We know this when we are on a train (assumed to be traveling at a constant speed) and everything we do on the train is basically the same as when we are on the ground. Or take the second law of thermodynamics, as championed by Clausius: no process is possible that transfers heat from a cold to a hot object without any other effect. This is why we all use fridges and understand the need to pay for the electricity to power them. Both of these principles rule out many things that might, at least logically and at first sight, seem perfectly sensible.

The ultimate laws would prohibit *all* the impossibilities, so that whatever is left must be reality's entire set of possibilities.

This kind of approach to physics sounds a bit like Michelangelo's supposed reply to a passerby who marveled at the skill and ingenuity that went into making the statue of David. "Oh, there is nothing to it," Michelangelo said. "I just took a large block of stone and removed the superfluous bits."

But how do we remove the superfluous bits from the "mathematical" block comprising all imaginable universes so that all we are left with is our own universe? Once more, quantum may show us the way.

QUANTUM ALL THE WAY DOWN?

We've already encountered the philosophy of infinite regress when we talked about time in Chapter 5. But my main point here will not be about philosophy. As far as I am concerned, physics is the queen and philosophy just one of its many servants. I bring this philosophy back up to tell you about an interesting speculation about the next physical theory that deserves more attention.

Quantum interactions between different subsystems, such as an atom and the electromagnetic field, are always characterized by Hamiltonians. In these instances, the Hamiltonian is a function of both q-numbers (corresponding to, say, the subsystems' position and momentum) and c-numbers (corresponding to, say, the subsystems' various charges, as well as the speed of light).

This situation may not seem satisfactory, as we might like all the entities in our theory to be of the same kind. Shouldn't the quantum Hamiltonian contain only quantum things—all q-numbers and no c-numbers? My colleague Deutsch wrote a paper a while back—for Bryce DeWitt's sixtieth-birthday collection—contemplating what it would mean to eradicate all c-numbers from quantum physics. What would such a world look like?

Another colleague of mine, Tomek Paterek, put the question somewhat differently. He said that if the interaction between an atom and the electromagnetic field contains quantum numbers other than the ones pertaining to the two systems (like the dipole and the electric field q-numbers), then there ought to be a mediator in the form of a *third* quantum system that actually "facilitates" the interaction between the atom and light.

So when an excited atom de-excites and emits a photon, Paterek would say that there should be another quantum system that couples to both the atom and the electromagnetic field. In other words, quantum electrodynamics would have to be modified, since in addition to electrons and photons there would now be more quantum things that ought to be taken into account.

Both Deutsch's and Paterek's ideas can be tested (i.e., they are not just philosophy), as they would lead to novel effects that do not exist in the current model.

Which brings me to the title of this subsection. There is a famous story involving Bertrand Russell in which he gave a public lecture explaining how the Earth revolves around the Sun and how the force of gravity keeps the whole universe in a state of balanced motion. At the end of the lecture, an old lady challenged him by stating that she thought the Earth was supported by a giant turtle instead. Russell, thinking that he would get the better of her, asked: "But what does this turtle stand on, madam?" To which she replied simply: "It's turtles all the way down!"

Now, I'd like to outline a picture of the universe in which it's quantum physics all the way down. I will take a different road than Deutsch and Paterek but will effectively arrive at the same conclusion.

When we upgrade the classical Hamilton-Jacobi equation[*] for quantum physics, we obtain the Schrödinger equation. This upgrading procedure is known as the first quantization, in which some (but not all) c-numbers are converted to q-numbers. The first quantization forces allowed some energies

---

[*] This is a way of expressing Newton's laws of motion in which the motion of a system can be thought of as a classical wave.

of the system to become discrete (such as the energy of the low-est electron orbital in the hydrogen atom) and therefore different from their continuous classical counterparts. However, these energies are still c-numbers.

Now, there is something called the second quantization. This takes the energies from the first quantization and converts them into q-numbers. This procedure involves introducing an extra complexity to each energy. In other words, now we have a superposition of one particle with a given energy, two particles with that very same energy in total, and so on. When this second quantization is implemented, we obtain quantum field theory, the most accurate description of all known phenomena in physics.

But why stop there? Why not a third quantization? This question has been asked by many researchers, perhaps first by Yoichiro Nambu (who won the Nobel Prize for spontaneous symmetry breaking) more than seventy years ago. What would be the physical meaning of a third quantization? It would imply that photons are not the ultimate fundamental units of light, but that they consist of some underlying quanta, which James Franson (the latest in the line of third-quantizers) calls "oscillatons" (because photons are just excitations of the simple harmonic oscillator). So each photon could, in fact, be made up of superpositions of different numbers of oscillatons!

The existence of such oscillatons could in principle be tested for, just as we can test for the existence of photons. We do not even need to detect oscillatons directly—it would suffice to detect some deviation in the current laws of quantum electrodynamics that could be explained by the existence of oscillatons (photons, too, do not have to be detected directly;

instead we can observe atomic oscillations that can only be explained by the existence of photons).

Extrapolating, we could even imagine fourth, fifth, and sixth (and so on) quantizations, each leading to the existence of more and more particles, but never uncovering the ultimate fundamental constituents. Maybe reality always contains yet another, deeper level for us to uncover.

Philosophers frequently object to the idea of an infinite regress, since they believe it to be a bad way of explaining anything. The story of Russell and the old lady is meant to illustrate exactly that point. But Nature does not care about philosophers. All the things that we call paradoxical in our philosophy are simply mismatches between what we believe reality *ought* to be and what it actually really *is*.

It is certainly possible that the universe does not have any ultimate building blocks out of which everything else is made. And maybe this is good news for us physicists (and all other truth-seekers), as it means we would never run out of things to discover. But there is always the question of how quickly we can leap to new ideas. There could be some natural limits to our ability to understand and perceive the reality around us. I want to contemplate what would happen if we applied all the ideas so far to explaining our own mind and, more interestingly, to enhancing its current capabilities. This is the last bit of the arsenal I will uncover, and it may lead us to a transformative discovery.

THE PORTAL TO QUANTUM PERCEPTION

The main thesis of this book has been that we need to understand the quantum world in order to discover the next revolution in physics. More specifically, we need to take seriously the idea that the world is even more quantum than most physicists realize. However, all the machinery I have examined so far (mathematics, language, analogies, philosophy, principles, and even "extreme" quantization) may not be sufficient to embrace the quantum universe fully. I think we need to contemplate the possibility of a more cerebral approach.

The clues come from considering this question: why have scientists for so long resisted the view that everything really behaves quantum mechanically, from the smallest particle to macroscopic objects to the entire universe? A natural explanation is that our senses have evolved in the macroscopic world, which is—to a good approximation—governed by the laws of classical physics. In the macroscopic world, the intuitive rules of logic are those laid down by George Boole: either something exists out there or it doesn't; either it is located at one place or it is located at another. Not only that, but we are used to the world in which objects move at small velocities compared to the speed of light. This gives us another false intuition that relativity contradicts. Our gut tells us that space and time exist independently of each other and of the rest of the stuff in the universe, but, as we've seen, physics tells us otherwise.

It is therefore the fault of our physiology and psychology that our natural mental model of the world is essentially Newtonian, rather than Einsteinian or quantum mechanical. Even when we professional physicists describe what happens in quantum physics as well as relativity, Newtonian elements still subconsciously creep into our explanations. One could even say that it is impossible to communicate non-Newtonian ideas without using our intuitive, Newtonian notions!

The picture that I have been promoting is that ultimate reality consists of q-numbers, which describe all the fields in the universe and possibly even spacetime itself. The principles of quantum information describe the relationship between different q-numbers, including the all-important phenomenon of entanglement. Counterintuitively, entanglement is also at the heart of the emergence of the classical world of definitive states and outcomes. Now, the amount of entanglement between two or more systems could be any number between zero and its maximum. It is only at the maximal point that we have the "classical" scenario of Wigner's Friend. All other entanglements correspond to partial correlations between q-numbers.

Is there a way for us to grasp this underlying reality without relying on abstractions and mathematics? Could we actually *experience* the world of q-numbers and quantum phenomena more directly? Among other things, this amazing achievement would end the need to use classical analogies in order to talk about quantum phenomena. Imagine perceiving q-numbers directly, feeling them as readily as we sense heat or see colors.

The idea that we only perceive a small portion of reality is, of course, not new. The poet William Blake famously said in 1790: "If the doors of perception were cleansed every thing would appear to man as it is, infinite. For man has closed himself up, till he sees all things thro' narrow chinks of his cavern."

Aldous Huxley wrote a book whose title he borrowed from Blake, *The Doors of Perception*, in which he advocated the use of chemical substances in order to enhance the human capabilities of seeing a much larger reality than our natural senses give us access to. This ushered in the new hippie era (Jim Morrison and his band the Doors come to mind), in which substance abuse became one of the defining features of the young generation who wanted to change the world. Whether drugs open us up to a larger reality or distort our perception of actual reality has been a subject of many a debate.

Be that as it may, I'd like to claim that the physics I've been describing in this book is a far more potent way to lead us beyond the "classical" doors of perception. And this, I think, would be the ultimate transformation instigated by the new physics, to a view that regards the world as consisting of q-numbers. It would not just be an intellectual revolution due to the discovery of a new theory. It would not just lead to new technologies, as we've always seen happen in the past whenever new physics was discovered. Rather, it would take us to an entirely new mode in which our species could experience reality. The first steps can already be taken within our current understanding, but what follows beyond could only make our journey even more exciting.

Here is what I have in mind. First of all, it is clear that our senses are limited to a tiny set of impulses from the outside world. For instance, we can see only a narrow range of frequencies of light and hear only a narrow range of sounds. The same goes for our other senses. But even more stringent than these limitations is that our senses are confined entirely to *classical* inputs and outputs. So while our senses could be enhanced to detect, say, a wider range of light frequencies and sound amplitudes, we would merely be broadening our experience of *classical* reality. Perhaps this would satisfy Blake and Huxley, but in terms of qualia—how we experience things—this would not change our perceptions in a fundamental way.

The reason our experience is classical is that our brain activity operates according to Boolean logic—a neuron either fires or it doesn't. This results in a signal that either is strong enough to stimulate awareness or isn't. Such a firing does not exist in a superposition of states. After all, quantum physics says that we cannot measure or experience all the properties of a system when we engage with it.

This brings me to an exciting possibility based on what is called a "weak measurement" in quantum physics. A weak measurement is an interaction between two systems during which they entangle with each other very weakly, such that the q-number that is being measured is only weakly disturbed.* In other words, it almost stays intact!

An apt example of this is a measurement of an electron that is in a superposition of two different locations. It's apt because our perception is underpinned by electrons moving down and between the neurons in our brain. One way of measuring the position of the electron is to bring another electron close enough to it so that the second electron moves one way if the first electron is in one place and another way if the first electron is in the other place. However, if the second electron moves only a bit (that is, if it entangles itself with the first electron only a bit), then the measurement is weak.

The downside is that the information obtained in weak measurements is also, well, weak. In other words, although the q-reality is not disturbed much, we also have not learned much. In our electron example, the second electron's position does not conclusively tell us about the position of the first electron. But what if we made simultaneous weak measurements of things that cannot be measured simultaneously, like

---

* This is why I said that the many-worlds interpretation is not the ultimate picture of the fully quantum universe, since it is based exclusively on strong measurements (i.e., maximal entanglements). I instead insist on the alternative, more general view that "everything is a q-wave."

the position and the velocity of the electron? Although it is physically possible to measure many q-numbers at once very weakly, it is not what our brain does. Our brain presumably makes highly entangling, definitive measurements that give us maximal information. Therefore, we would have to redesign our neurocircuitry in order to accommodate such processes. To successfully engineer our brain so that it could simultaneously and weakly measure many q-numbers, we would need to integrate specially designed microchips that carry out these quantum measurements in conjunction with the existent machinery of our brain.

What would this feel like? We can't yet know, since we would first need to understand the deeper question of how any of our feelings arise in the first place. But it is quite possible that experiencing q-numbers would transcend any other experience we've been able to engineer in our history, all while offering an experience that corresponds more closely to fundamental reality than drugs could ever offer.

I think this is the biggest transformation that the research I have been describing is capable of initiating. Not only could our quantumly augmented perception enable all of us to have a direct understanding of the underlying physical reality and experience the emotions related to this, but it is quite possible that this newly acquired quantum intuition would in itself be of huge advantage for our survival. Maybe, just maybe, it would allow us to build a better and more robust society than anything we've been able to master so far.

Now, there's a thought for the twenty-first century. Turn on, tune in, and follow me on this adventure to reach the portals into a new reality.

ACKNOWLEDGMENTS

This book is the result of a journey I embarked on about ten years ago when it occurred to me that quantum information won't just improve our technological capabilities and understanding of quantum mechanics but also lead us to an entirely new theory of physics. I am still on that road, but a couple of years ago it became clear that there's already enough exciting stuff to communicate. We are now knocking on the portals of the future of physics, and I am confident we will bust through them sometime in the coming decade or two. As Carl Sagan said, "When you are in love, you want to tell the world," but it's not only this that compelled me to write. The book is a call to arms from the battlefront to explore wholeheartedly the new emerging reality. The best way to predict the future, after all, is to create it yourself.

I might be on a road less traveled, but I have a small and awesome band of brothers and sisters in arms. I've benefitted a great deal from their insights, discussions, and continuous support. Without Charles Bennett, David Deutsch, Artur Ekert, Mile Gu, Chiara Marletto, Tomek Paterek, Benjamin Schumacher, William Wootters, and Anton Zeilinger, none of this would have been as fun and fulfilling as it was.

ACKNOWLEDGMENTS

Special thanks go to Chiara Marletto and Logan Chipkin for all their comments and suggestions. Logan's and Sue Warga's dedicated line editing is greatly appreciated. Finally, TJ Kelleher of Basic Books has been crucial for his guidance and encouragement. His detailed readings of my manuscript and perceptive editorial insights have immensely improved my presentation.

Oxford 2025

FURTHER READING

Olivier Darrigol, *From c-Numbers to q-Numbers: The Classical Analogy in the History of Quantum Theory* (University of California Press, 1992).

B. L. Van Der Waerden, *Sources of Quantum Mechanics* (Dover Publications, 2007).

Ian J. R. Aitchison, David A. MacManus, and Thomas M. Snyder, "Understanding Heisenberg's 'Magical' Paper of July 1925: A New Look at the Calculational Details," *American Journal of Physics* 72, no. 11 (2004): 1370–1379.

Y. Frenkel, *Wave Mechanics: Advanced General Theory* (Clarendon Press, 1934).

Michel Bitbol, *Schrödinger's Philosophy of Quantum Mechanics* (Springer Netherlands, 1996).

Charles W. Misner, Kip S. Thorne, and John Archibald Wheeler, *Gravitation* (Princeton University Press, 2017).

Arthur Koestler, *The Sleepwalkers: A History of Man's Changing Vision of the Universe* (Compass, 1989).

J. C. Taylor, *Hidden Unity in Nature's Laws* (Cambridge University Press, 2001).

Abraham Pais, *Subtle Is the Lord: The Science and the Life of Albert Einstein* (Oxford University Press, 2005).

Vlatko Vedral, *Decoding Reality: The Universe as Quantum Information* (Oxford University Press, 2018).

Sunny Y. Auyang, *How Is Quantum Field Theory Possible?* (Oxford University Press, 1995).

Jennifer Coopersmith, *The Lazy Universe* (Oxford University Press, 2017).

J. W. Dunne, *An Experiment with Time* (Dover Publications, 2019).

Yakir Aharonov and Daniel Rohrlich, *Quantum Paradoxes: Quantum Theory for the Perplexed* (Wiley, 2005).

FURTHER READING

Jean-Marc Lévy-Leblond and Françoise Balibar, *Quantics: Rudiments of Quantum Physics*, translated by S. Twareque Ali (Elsevier Science, 1990).

Bertrand Russell, *The ABC of Relativity* (Routledge Classics, 2009).

Tim Folger, "Quantum Gravity in the Lab," *Scientific American*, April 1, 2019.

Chiara Marletto and Vlatko Vedral, "Witness Gravity's Quantum Side in the Lab," *Nature* 547 (2017): 156–158.

Nick Huggett, Niels Linnemann, and Mike D. Schneider, *Quantum Gravity in a Laboratory?* (Cambridge University Press, 2023).

Osborne Reynolds, *On an Inversion of Ideas as to the Structure of the Universe (The Rede Lecture, June 10, 1902)* (Cambridge University Press, 1902).

Roger Penrose, "On Gravity's Role in Quantum State Reduction," *General Relativity and Gravity* 28 (1996): 581–600.

Hans C. Ohanian and Remo Ruffini, *Gravitation and Spacetime* (Cambridge University Press, 2013).

Michael Berry, *Principles of Cosmology and Gravitation* (Cambridge University Press, 1978).

E. T. Whittaker, *A History of the Theories of Aether and Electricity*, 2 vols. (Harper Torchbooks, 1960).

Michael Friedman, *Foundations of Space-Time Theories: Relativistic Physics and Philosophy of Science* (Princeton University Press, 2019).

Lee Smolin, *Three Roads to Quantum Gravity* (Basic Books, 2017).

Richard Feynman, "The Reason for Antiparticles," in *Elementary Particles and the Laws of Physics*, ed. Richard Feynman and Steven Weinberg (Cambridge University Press, 1987).

Charles G. Darwin, *The New Conceptions of Matter* (Bell & Sons, 1932).

C. Marletto and V. Vedral, "Quantum-Information Methods for Quantum Gravity Laboratory-Based Tests," *Reviews of Modern Physics* 97 (2025): 015006, https://arxiv.org/abs/2410.07262.

John Archibald Wheeler and Wojciech Hubert Zurek, eds., *Quantum Theory and Measurement* (Princeton University Press, 1983).

John D. Barrow and Frank J. Tipler, *The Anthropic Cosmological Principle* (Oxford University Press, 1988).

Richard H. Beyler, "Targeting the Organism: The Scientific and Cultural Context of Pascual Jordan's Quantum Biology, 1932–1947," *Isis* 87, no. 2 (1996): 248–273.

FURTHER READING

John L. Heilbron, "The Earliest Missionaries of the Copenhagen Spirit," in *Science in Reflection: The Israel Colloquium: Studies in History, Philosophy, and Sociology of Science, Volume 3*, ed. Edna Ullmann-Margali (Kluwer Academic, 1988).

Michel Blay, "Bohr et la complémentarité," *Revue d'Histoire des Sciences* 38, nos. 3–4 (1985): 195–230.

Art Hobson, *Tales of the Quantum: Understanding Physics' Most Fundamental Theory* (Oxford University Press, 2017).

Erwin Schrödinger, *What Is Life?* (Cambridge University Press, 1944).

Vlatko Vedral, *From Micro to Macro* (World Scientific, 2018).

Lisa Grossman, "In the Blink of Bird's Eye, a Model for Quantum Navigation," *Wired*, January 27, 2011, www.wired.com/2011/01/quantum-birds.

John Baez, "The Story of Nth Quantization," March 14, 2016, https://math.ucr.edu/home/baez/nth_quantization.html.

Yoichiro Nambu, "On the Method of the Third Quantization," *Progress of Theoretical Physics* 4, no. 3 (1949): 331–346.

J. D. Franson, "Third Quantization of the Electromagnetic Field," *Physical Review A* 104 (2021): 063702.

Aldous Huxley, *The Doors of Perception and Heaven and Hell* (Harper & Brothers, 1956).

INDEX

INDEX